Historic Orchard and Fruit Tree Stabilization Handbook

National Park Service

Pacific West Region, Cultural Landscapes Program

with

California Department of Parks and Recreation

Archaeology, History and Museums Division

2012

The authors wish to thank those who contributed to the development and completion of this document, including:

Project Team

Susan Dolan, Principal Author, Historical Landscape Architect (NPS, PWR)

Cortney Cain Gjesfjeld, Co-Author, Historical Landscape Architect (NPS, PWR)

Cathy Gilbert, NPS Project Manager, Acting Lead, Cultural Landscapes Program (NPS, PWR)

Jan Wooley, CDPR Project Manager, State Historian III, Archaeology, History and Museums Division (CDPR)

Other Consultants, Reviewers and Advisors

John Fraser, State Historian I, Archaeology, History and Museums Division (CDPR)

Marianne Hurley, State Historian II, Diablo Vista, Marin and Russian River Districts (CDPR)

Kathleen Kennedy, State Historian II, Archaeology, History and Museums Division (CDPR)

Kimball Koch, Historical Landscape Architect (Yosemite National Park)

CONTENTS

Stabilization Actions

Appendices

Glossary

LIST OF FIGURES

Figure 13: Example of an orchard management plan, NPS document for Manzanar National Historic Site.

Figure 14: A historic pear orchard before stabilization at Manzanar National Historic Site in Inyo County.

Figure 15: Five years later, the historic pear orchard at Manzanar National Historic Site has been stabilized and is responding well to stable conditions.

Figure 16: Historic pear orchard at Jack London State Historic Park is threatened by encroaching vegetation, creating unstable conditions.

Figure 17: A structurally unstable historic pear tree at Manzanar National Historic Site was toppled by wind-throw.

Figure 18: Presence of deadwood and downed debris are sources of infection among these historic almond trees.

Figure 19: Diagram showing the spectrum of natural longevity in fruit trees is species-dependent, but is also affected by type of root system.

Figure 20: NPS and CDPR staff use the fruit tree condition assessment field form to determine this historic pear tree is in fair condition.

Figure 21: Recommended fruit tree condition assessment field form.

Figure 22: Condition assessment field form "Zone 0" is the orchard floor immediately outside the tree canopy. In this apple orchard in the North Coast Redwoods District, Zone 0 is unstable, due to colonizing blackberries.

Figure 23: Condition assessment field form "Zone 1" is the orchard floor within the dripline of the canopy. At this pear tree, Zone 1 is stable with no encroaching vegetation.

Figure 24: Condition assessment field form "Zone 2" is the base of the trunk at the point of attachment to the roots. At this pear tree, Zone 2 is unstable, due to the presence of a rodent burrow and root suckers.

Figure 25: Condition assessment field form "Zone 3" is the tree trunk. At this pear tree, Zone 3 is unstable, due to the presence of multiple shoots (epicormic growth) on the trunk, and root suckers.

Figure 26: Condition assessment field form "Zone 4" is the canopy of the tree. At this pear tree, Zone 4 is unstable due to severe dieback, attached deadwood, and sparse foliage.

Figure 27: Condition assessment field form "Zone 5" is the area above or around the tree canopy. At this plum tree, Zone 5 is unstable, due to the over-shading presence of a Live oak tree.

Figure 28: Sample segment of an orchard site map, showing tree identification number, location, and condition classes (green = fair, yellow = poor, cross = dead).

Figure 29: CDPR and NPS staff celebrates the successful stabilization of a historic quince tree at Jack London State Historic Park.

Figure 30: Two site maps of the historic orchard system at Jack London State Historic Park showing two approaches to stabilization. The top map shows the area to be fully stabilized, and the lower map shows a smaller scope: partial stabilization.

Figure 31: Sample fruit tree stabilization record, tracking the date and type of stabilization actions by tree.

Figure 32: Aerial photo depicting stable historic orchard conditions at the Buckner Orchard in North Cascades National Park Service Complex, WA.

Figure 33: Diagram distinguishing a seedling fruit tree from a grafted fruit tree, and indicating the scion and rootstock components of a grafted tree.

Figure 34: A historic pear tree at La Purisima Mission State Historic Park bears an abundance of fruit.

Figure 35: Stable orchard floor conditions in the historic Buckner Orchard in North Cascades National Park Service Complex.

Figure 36: Examples of a walk-behind brush hog (upper photo) and an ATV-pulled brush hog (lower photo), to remove woody brush on the orchard floor.

Figure 37: A common need in neglected orchards is root sucker removal from the base of trees.

Figure 38: Historic pear tree lower trunk at Manzanar National Historic Site after sucker removal.

Figure 39: NPS staff member at John Muir National Historic Site demonstrates the use of a rider mower to mow within a low-headed grape vineyard, suitable also for orchard floors.

Figure 40: Diagram showing tree conditions before and after cleaning activities.

Figure 41: Basic hand tools and personal protective equipment for cleaning fruit trees.

Figure 42: Diagram showing the recommended cut locations for the removal of a dead scaffold limb and a dead co-dominant scaffold limb.

Figure 43: Diagram showing the Target Pruning Method for deadwood removal, where the cut is made immediately outside the branch bark collar.

Figure 44: Photo showing a poor pruning cut where a limb was removed by cutting inside the branch bark collar, i.e., too deep.

Figure 45: A toppled historic pear tree at Fort Ross State Historic Park is still alive, but severely compromised in condition.

Figure 46: Diagram of cabling to prevent two scaffolds splitting apart.

Figure 47: Photo of a prop supporting a leaning apple tree, preventing the tree from becoming severed at the roots.

Figure 48: Examples of braces: vertical braces support a leaning tree (upper photo) and a horizontal brace prevents a tree from splitting apart.

Figure 49: An appropriately-scaled rider mower for maintaining the orchard floor in-between the rows and columns of trees in an orchard.

Figure 50: A stable historic orchard in Olympic National Park has received deadwooding, aerating, and then mulch within the driplines of trees.

Introduction

This handbook is designed to serve as a guide for stabilizing the condition of potentially historic or known historic orchards and fruit trees within the California state parks system. It is intended for use by park managers, resources managers and maintenance staff, partners and volunteers that are responsible for the planning, management or preservation maintenance of orchards or fruit trees. The handbook integrates the concepts of historic preservation and orchard horticulture to guide efforts to prevent the further deterioration in condition of potentially or known significant biotic cultural resources.

The need for this handbook was recognized by the California Department of Parks and Recreation (CDPR) Archaeology, History and Museums (AH&M) Division. Orchard remnants and fruit trees were known to exist in state parks throughout California, with ties to the state's rich land settlement and agricultural history. Most of these orchards or fruit trees were unevaluated and unmaintained, however, and the State was at risk of losing potentially significant cultural resources and the information they contained. The AH&M Division identified the need for technical guidance and staff training to arrest deterioration until planning could be conducted for the future treatment of these resources.

Figure 1: A visitor enjoys a peaceful walk in the historic olive orchard at Folsom Lake State Recreation Area (CDPR, 2010).

This handbook was prepared by the National Park Service (NPS) in partnership with CDPR AH&M Division through a cooperative agreement. The NPS is the nation's lead historic preservation agency, directed by the National Historic Preservation Act of 1966 to participate in a national preservation partnership with American Indian tribes, states, local governments, nonprofit organizations and historic property owners. The NPS develops standards and guidelines for historic preservation and offers hands-on guidance for preservation partners. The NPS Cultural Landscapes Program has developed registration requirements for listing orchards and fruit trees in the National Register of Historic Places.

In 2009, the NPS published a national historic context of orchards in the United States, entitled *A Fruitful Legacy: A Historic Context of Orchards in the United States, with Technical Information for Registering Orchards in the National Register of Historic Places.* This document highlights the significance of orchards and fruit trees in America's settlement and economic development history, and identifies significant events, trends and characteristics in the evolution of orchards since the 1600s. The document provides the basis for the analysis and evaluation of orchards and fruit trees as historic properties, using the National Register criteria. This Stabilization Handbook is intended to compliment *A Fruitful Legacy* to guide subsequent actions leading to the preservation of orchards and fruit trees that are known to be, or are potential cultural resources.

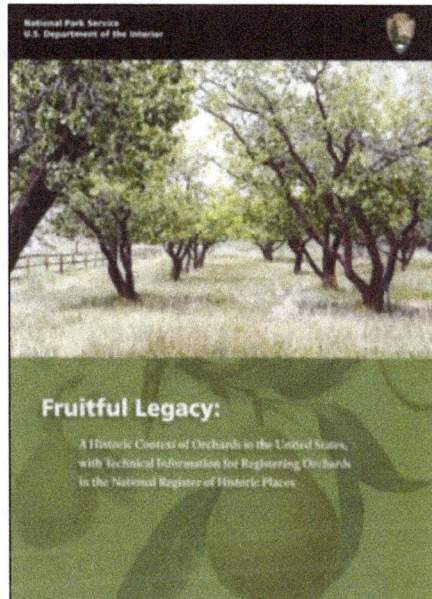

National Park Service
U.S. Department of the Interior

Fruitful Legacy:

A Historic Context of Orchards in the United States, with Technical Information for Registering Orchards in the National Register of Historic Places

Figure 2: The NPS publication *Fruitful Legacy* serves as a historic context of orchards and provides guidance for National Register listing (NPS, 2009).

The needs of CDPR for information and training on orchard assessment and stabilization were addressed by the AH&M Division with three services provided by the NPS through the agreement: a CDPR orchard/fruit tree survey; a historic orchard stabilization workshop; and a stabilization handbook. The system-wide survey was conducted in 2008, yielding results that at a minimum of 44 state parks have potentially historic orchards or fruit trees (see the Appendices for detailed information), and the workshop was held in March 2010 at Jack London State Historic Park, attended by cultural resources and maintenance staff and volunteers. The workshop was filmed and developed into a training video by AH&M Division, entitled "Historic Orchard Assessment". This handbook addresses the range of orchards and fruit trees identified through the system-wide survey, and provides guidance on maintenance issues identified by CDPR staff during the workshop.

Figure 3: CDPR staff and partners at the 2010 orchard workshop (CDPR, 2010).

Historic Context

Introduction

A 2008 system-wide survey of California State Parks revealed that a minimum of 44 parks, or at least 15 percent of the entire system, have potentially historic orchards or fruit trees. This is a high percentage, considering the system contains the most diverse natural and cultural heritage holdings of any state agency in the nation. California state parks include such diverse units as underwater preserves, redwood, rhododendron and wildlife reserves, state beaches, recreation areas, lighthouses, caverns, ghost towns and off-highway vehicle parks. The survey results indicated a widespread geographic distribution of orchards and fruit trees throughout the system, highlighting the importance of orchard fruit growing in California land settlement and development history. A large number of orchard fruit species was also revealed, reflecting California's place in the nation as home to the greatest diversity of fruit and nut crops. In 2008, the following orchard fruit species were reported in California State Parks: apple, apricot, cherry, fig, lemon, mulberry, nectarine, olive, orange, peach, pear, pecan, persimmon, pistachio, plum, pomegranate, prune, quince and walnut.

History of California Fruit Tree Growing

The history of orchards in California can be traced to the introduction of non-native fruits by Spanish missionaries between 1769 and 1823. Before this time in the lands to become California, Native Americans for thousands of years intensively managed their environment and used native plants for food, medicine and crafts. According to Kat Anderson, ethnoecologist at the University of California in Davis, Native Americans enhanced the production of useful plants with fire, obsidian knives, and pointed digging sticks. Fruit tree plantations in the form of orchards did not exist, though the fruits of many native trees were abundantly harvested, including Choke cherry *(Prunus virginiana)*, American plum *(Prunus americana)*, Pacific crabapple *(Malus fusca)*, Mexican elderberry *(Sambucus mexicana)*, Black walnut *(Juglans hindsii)*, Valley oak *(Quercus lobata)*, California black oak *(Q. kelloggii)* and California fan palm *(Washingtonia filifera)*. The abundance of plant and animal food in these lands supported one of the densest aboriginal populations in the country. Almost all temperate orchard fruit species originated in the lands of Asia or Europe, however, and were absent from the New World until their introduction by European settlers.

From 1495 onwards, Spanish citizens "of good standing" and later, missionaries and priests, were permitted to emigrate to North, Central and South America, including Baja California. The establishment of agriculture in the New World was specifically encouraged by the Spanish Monarch, to settle and "civilize" lands claimed by the Crown. Emigrants carried young trees, cuttings and seeds of orchard fruits with them on their oceanic voyages. Mexico City would become a major supply hub of horticultural stocks in the New World, serving as a dissemination point for imported crop plants. Spanish settlers planted fruit trees in Hispaniola (the Dominican Republic and Haiti), Venezuela and throughout Mexico, including Baja California, where 15 missions with orchards were established by Jesuit priests. In the 1760s, the King of Spain directed his troops and priests to take possession of the lands of Alta California. Father Junipero Serra was the missionary assigned to lead the effort, and between 1769 and 1823, 21 Franciscan missions were established along the California coast, from Mission San Diego de Alcalá in San Diego northward to Mission San Francisco Solano, in Sonoma. Most missions would be planted with gardens, orchards and agricultural fields.

While the principal purpose of the missions was to convert American Indians to the Christian faith and lay claim to the Spanish territory, their primary activities were associated with subsistence agriculture. Each mission needed to become self-sufficient as rapidly as possible, as the existing means of supply was inadequate to sustain a colony of any size. Barley, maize and wheat were among the most common crops grown, and olive, orange, apple, peach, pear, plum, fig and pomegranate were the most common orchard fruits. The orchards, fields and gardens were tended by colonists or converted Native Americans, who, once baptized, were no longer free persons. In 1806, a total of 20,355 Native Americans were attached to the California missions, where they labored under the strict observance of priests and mission soldiers. The first recorded orchard planting was a peach orchard at Mission San Gabriel Arcángel, in San Gabriel in 1769. The peach was probably selected for its species' fast-growing and youthful-bearing characteristics. By 1792, British Navy Captain George Vancouver noted that the orchard of the Mission San Buenaventura in Ventura contained peach, apricot, apple, pear, fig, plum and pomegranate trees, along with grapevines and banana plants. In 1803, Father Fermin Lasuén, successor to Father Serra, reported that the missionaries at Mission San Diego had pressed olives from the mission orchard, producing their first olive oil.

Some mission orchard trees were cultivated varieties grafted as young trees onto seedling rootstocks. Other fruit trees were un-grafted seedlings of no variety that were started from seed or rooted cuttings. Peach and plum could be grown from seed to produce edible fruits, while fig and olive trees could be grown from rooted cuttings. Distinct from most temperate orchard fruit trees, fig and olive varieties can be reproduced from rooted cuttings. The varieties of most orchard trees can only be reproduced through the process of grafting a scion to a rootstock. All of the orchard trees would have had tall trunks (more than 5 feet tall), as their lower limbs would have been browsed off by grazing animals. The mission orchards were irrigated through extensive water supply systems. Stone zanjas (aqueducts) sometimes spanning miles, brought fresh water from a nearby river or a spring to the mission fields. Baked clay pipes, joined together with lime mortar or bitumen, deposited water into large cisterns and gravity-fed fountains, which emptied into irrigation ditches between the rows and columns of orchard trees.

After Mexico achieved independence from Spain through the Mexican War of Independence, the Mexican government gained ownership of the lands of California and closed down the missions, confiscating the land. The Franciscan priests were forced to abandon their mission lands and return to Spain. Some Spanish colonists remained behind in presidios, pueblos or on some 30 ranchos granted by the Spanish government to encourage agriculture and industry. During Mexico's control of California between 1823 and 1848, another 770 ranchos were granted by the Mexican Government to Mexican citizens.

Figure 4: Pear tree with tall-trunk form typical of the 1820s at La Purisima Mission State Historic Park (CDPR, 2010).

The Spanish and Mexican ranchos were limited to a maximum size of 11 square leagues, or 48,712.4 acres. They were located on fertile lands and were extensively cultivated with orchards, vineyards, and crops or grazed with livestock. The land of Jack London State Historic Park was first privately owned as one such rancho. This parcel of land was part of the Petaluma, Aqua Caliente Grant that was given to General M.G. Vallejo by the Mexican Government in 1834, and later became part of the Sonoma Developmental Center.

At the end of the Mexican-American War of 1846 to 1848, Mexico conceded the lands of California, Texas, Arizona and New Mexico to the United States. The 1848 Treaty of Guadalupe Hidalgo established the rights of Mexicans to their rancho land titles within the conquered territories. Some ranchos were owned by naturalized Mexican citizens, such as William Wolfskill, an American fur trapper who had emigrated during the 1830s. Wolfskill's rancho was located near the pueblo of Los Angeles, where he planted an extensive orchard with seeds and cuttings from the mission lands, including apple, pear, orange and olive. Wolfskill is considered to be the first Californian to grow oranges commercially. Other American immigrants followed suit during the Mexican period, purchasing land and planting orchards to sell fruit. Carefully packed fruits were shipped by sea from Los Angeles to markets in San Francisco.

The California Gold Rush began on January 24, 1848, when gold was found by James W. Marshall at Sutter's Mill in Coloma. News of the discovery brought approximately 300,000 people to California from the United States, Latin America, Europe and China, transforming pueblos into boomtowns, and fueling the building of roads and the growth of agriculture to meet the needs of settlers. Immigrants expanded orchard growing from a localized phenomenon near mission lands to a modest commercial level. Many so-called "49ers" also turned their hands to horticulture. One such miner, William M. Stockton, laid claim to the former lands of San Gabriel Mission and rehabilitated the 75-year old orchards, including olive, orange, and pear trees. Stockton grafted the seedling rootstocks of the old Spanish pear trees with his own stock varieties, giving rise by 1853, to the first commercial pear orchard in California. Within 20 years of the Gold Rush, California had a booming orchard industry.

Cattle ranching was the first large-scale agricultural enterprise in California, however. The cattle, originally introduced by the Franciscans in the 18th-century, could thrive with relatively little care and were profitable from the Mexican period until the early 1860s. Bad weather led to the demise of the cattle hegemony, however. Floods struck the Sacramento Valley in 1861 and 62, where most cattle were ranched, leading to widespread losses. The floods were followed by several years of drought, sealing the fate of many cattle ranchers. Wheat farming superseded cattle ranching as the next large-scale agricultural enterprise in the state. Wheat had also been introduced by the Franciscans, who had shown the grain could flourish on mission lands. The now-arid valley plains made excellent wheat fields, having been well-manured by cattle. Engineers opened up the San Joaquin Valley to wheat farming by draining marshes and shallow lakes. Wheat crops yielded 20 years of profitability until the 1880s, when international competition rendered foreign exports of wheat unprofitable, and nutrient depletion of soils caused diminishing yields.

Figure 5: 1872 photo of Mission San Diego olive orchard, leased by orchardist Thomas Davies (courtesy of the San Diego Historical Society).

Gradually more and more farmers and ranchers turned to orchard fruit growing for higher profitability. Commercial fruit growing was not a totally new experience for many, as commercial orchards were well-established enterprises in European countries, and in the eastern, southern and midwestern American states from which many settlers came. Those without experience could acquire the knowledge from the large body of new horticultural literature, journalism and horticultural societies, including county Grange organizations. The "California Farmer" was the most important and long-lasting of the journals. By the 1850s, commercial nurseries had superseded mission orchards as the source of orchard stock. Pioneers in the California nursery business include William Wolfskill, B. D. Wilson, John Rowland and William Workman of Los Angeles, E.L. Beard and John Lewelling of Santa Clara County, and the Thompson Brothers of Napa and

Solano counties. California orchardists wasted no time in exploiting the completion of the transcontinental railroad in 1869. That year, the first shipment of California apples and pears was made via rail to eastern markets.

By the 1870s, the newly-created United States Department of Agriculture (USDA) vastly added to the pool of orchard horticulture literature in disseminating instructional orchard bulletins for growers. After 1887, the influence of the USDA was vastly expanded through the Hatch Act, in which the U.S. Congress authorized the funding of agricultural experiment stations throughout the country, to *"conduct original and other research, investigations and experiments bearing directly on and contributing to the establishment and maintenance of a permanent and effective agricultural industry"*. A system of agricultural experiment stations were created throughout California and the other states, serving as venues for field research and grower education in improved orchard horticulture with greater productivity.

In 1874, the USDA imported a variety of orange from Brazil and introduced it to Riverside, California. The variety was "Navel", a seedless orange that would become the catalyst for the development of a commercial citrus industry in California. Prior to the introduction of Navel, the orange trees grown by Wolfskill and other orchardists were seedlings, derived from the mission orchards. "Mission oranges" were smaller-fruited with soft flesh and many seeds; in contrast, the Navel variety was large, firm, juicy and highly flavored.

The variety had other favorable characteristics for commercial growing, such as high yielding and early-bearing of young trees. A southern California "citrus belt" developed rapidly in the 1870s after the first experimental plantings in Riverside. Within two decades, commercial orange orchards stretched eastward from Pasadena to Redlands, beneath the foothills of the San Gabriel and San Bernardino mountains. The success of the Navel orange in Riverside led to the establishment of the University of California Citrus Experiment Station in 1907, the founding unit of the University of California, Riverside.

Contrary to a general decline in number throughout the country, the Pacific States added orchards and increased their number of fruit trees planted between the 1880s and World War II. While orchardists in other states were abandoning orchards due to burgeoning disease infestations and migration to urban industrial centers, newly settled lands in California were being planted as commercial orchards. The development of transportation and irrigation systems by the Federal government buoyed orchard planting in California, along with technological advancements in canning. The invention of cold storage in the 1890s was another boost to the California orchard industry, which depended on the lengthy haulage of fruits to eastern markets for commercial viability (at the turn of the 20th- century, 90% of the U.S. population lived east of the Mississippi River).

Figure 6: One of the original Navel Orange trees introduced to Riverside in 1874 and transplanted by President T. Roosevelt in 1903 (University of Chicago, 1903).

The regional development of commercial orange orchards in southern California marked the onset of a pattern in regionalism in orchard fruit growing throughout the state. Six main orchard regions would develop, based on the climatic influences of elevation, marine or fog influence, wind pattern, rainfall, slope, frost-free days, average temperature and temperature extremes. The six regions are the San Joaquin Valley, the Sacramento Valley, the Central Coast and North Coast, the Sierra Nevada Foothills and Southern California. The following table summarizes the regional distribution of orchard fruit growing in California that had evolved by the late 19th-century.

Figure 7: The following table shows the regional distribution of California orchards, by species (NPS, 2011).

Regional Distribution of California Orchards, by Species			
Common Name	Species	Dominant Regions	General Climatic Requirements
Almond	*Prunus dulcis*	Central Coast, Sacramento and San Joaquin Valleys, Southern California (coastal)	Absence of frost during early spring blossoming; well-drained soils, hot summers for nut maturation
Apple	*Malus domestica*	Northern Coast, Central Coast, Sierra Nevada Foothills	Cold winter; full sun exposure; ample water during fruit development
Apricot	*Prunus armeniaca*	Central Coast	Cold winter; absence of frost during early spring blossoming; dry spring resists diseases
Cherry (Sweet)	*Prunus avium*	Central Coast, Sacramento and San Joaquin Valleys,	Cold winter; well-drained soils; lack of extreme heat during fruit maturation
Cherry (Sour)	*Prunus cerasus*	All regions	Adaptable to most climates
Fig	*Ficus carica*	Central Coast	Mild winter; high heat during 2 crops: summer and fall; absence of frost during fall fruit development

Common Name (contd.)	Species	Dominant Regions	General Climatic Requirements
Lemon	*Citrus limone*	Central Coast, Southern California (coastal)	Mild winter; moderately hot summer
Mulberry	*Morus*	All regions	Adaptable to most climates
Nectarine	*Prunus persica*	All regions except Southern California	Cold winter; well-drained soil; ample water during fruit development
Olive	*Olea europaea*	Sacramento and San Joaquin Valleys, Southern California	Mild winter; hot summer; dry soils
Orange	*Citrus sinensis*	San Joaquin Valley, Southern California	Mild winter; hot summer
Peach	*Peach*	Central Coast, Sacramento and San Joaquin Valley, Sierra Nevada Foothills	Short winter chill requirement; hot summer; well drained soils; ample water during fruit development
Pear (European)	*Pyrus communis*	All regions except Southern California	Cold winter; hot summer; adaptable to different soil conditions

Common Name (contd.)	Species	Dominant Regions	General Climatic Requirements
Pecan	*Carya illinoensis*	San Joaquin Valley, Sierra Nevada Foothills, Southern CA	Very short winter chill requirement; very hot summer
Persimmon (Asian)	*Diospyros kaki*	San Joaquin Valley, Southern California	Mild winter; hot summer; adaptable to different soil conditions
Pistachio	*Pistachio vera*	Sacramento and San Joaquin Valleys	Cold winter; hot summer; dry soils
Plum (European) and Prune	*Prunus domestica*	All regions except Southern California	Cold winter; warm to hot summer; adaptable to different soil conditions
Plum (Japanese)	*Prunus salicina*	All regions except North Coast	Mild winter (short chill requirement); hot summer; adaptable to different soil conditions
Pomegranate	*Punica granatum*	Sacramento and San Joaquin Valleys, Southern California	Cold winter; hot summer; dry conditions during fruit development;
Quince	*Cydonia oblonga*	All regions	Adaptable to most climates
Walnut (English)	*Juglans regia*	Sacramento and San Joaquin Valleys	Absence of frosts during catkin blossoming; lack of extreme heat during nut development

Between the mid-19th-century and the turn of the 20th-century, orchard tree form changed in California and throughout the country from a five-feet or more tall trunk to a less than three-feet tall trunk, and tree shape changed from an un-pruned, relatively natural state to a pruned style with either a central leader or an open bowl shape. Also, fruit trees were planted with greater spacing in the orchard grid. The layout of apple, orange, walnut and olive orchards was expanded to 30 x 30 feet typical spacing. Tighter-spaced fruit trees such as pear, peach, plum and cherry were changed from a square to a rectangular arrangement, making room for machinery and increasing the yield of mature trees.

Figure 8: Diagram showing the evolution of fruit trees from the tall-trunk form in the 19th-century (left) to the short-headed form in the early 20th-century (right) (NPS, 2011).

A dramatic decrease in the number of varieties grown in California in the early 20th-century followed a national trend based on variety-selection for commercial characteristics. These characteristics promoted higher yields and greater durability of the harvested fruit. The number of varieties grown in commercial orchards was pared down from hundreds to tens. By World War II, most orchard fruit species were represented by just 10 widely grown commercial varieties, such as Red Delicious apple, Bartlett pear, Moorpark Elberta peach, Navel orange, Bing cherry, Hartley walnut, Nonpareil almond, Eureka lemon and Manzanillo olive. Thousands of fruit varieties became extinct or exceedingly rare, found only in old homestead orchards or on abandoned lands. John Muir, the renowned author and conservationist followed the contemporary trend in variety selection when he inherited his father-in-law's fruit ranch in the Alhambra Valley of Contra Costa County. From the 1880s until his death in 1914, Muir ran the fruit ranch profitably by adapting to modern methods.

Figure 9: John Muir's fruit ranch in the Alhambra Valley of Contra Costa County, showing widely spaced, low-headed fruit trees (NPS archives, 1915).

He vastly reduced the number of varieties that his father-in-law had grown, from hundreds in the 1860s to tens by the 1880s. The NPS preserves one of Muir's pear orchards at the site of his burial, in which Muir had re-grafted the pear trees in the 1890s with only Bartlett and Winter Nellis varieties.

The development of Red Delicious apple in the 1920s had an enormous influence on apple growing, resulting in greater profitability for the orchard industry, greater fashionability of red apples, greater ubiquity of a single variety, and further obsolescence of superseded varieties. Bartlett became the dominant pear variety, and California became the biggest producer. Curiously, while pear varieties generally require the presence of another variety to allow for cross pollination, Bartlett pear trees grown in the Sacramento and San Joaquin Valleys were known to be self-fruitful, setting crops of parthenocarpic fruits. Citrus, nut and olive orchards had a net increase in the number of varieties grown in the 1880 to World War II period. These industries were founded and established during this period as a result of the development of new varieties, such as Navel orange, Eureka lemon, Nonpareil almond, Manzanillo olive and Hartley walnut. California agricultural experiment stations and nurseries led the country in selecting varieties and developing horticultural techniques to propagate and transplant these tap-rooted trees. Once again, in this period these species were all low-headed, i.e., grown with short trunks on seedling rootstocks.

After World War II, the appearance of new orchards changed from widely spaced, full-size trees on seedling rootstocks, to dwarf, more closely-spaced trees. California orchards evolved along with the national trend toward the use of dwarf fruit trees for all new commercial and home orchards, along with the adoption of high density management systems, shrinking orchard spacing and tree size. The trend towards higher density, dwarf fruit tree orchards in the period from World War II to the present was fueled by the need to lower costs of production in an increasingly competitive marketplace. The discovery by European researchers that select dwarfing rootstocks could produce greater yields of higher quality fruit than seedling rootstocks led to the introduction of

Figure 10: A typical contemporary orchard of dwarf fruit trees planted at less than 5 feet apart within the rows, not exceeding 8 feet in mature height (NPS, 2008).

clonal dwarfing rootstocks into the United States, and their dissemination by the USDA after World War II. The trend toward more and more compact dwarf trees has continued for all hand-picked orchard species grown in California for the fresh fruit market. Trellising is also a common feature of contemporary orchards, where spindle trees cannot support themselves. Tree density has increased from 30 to 40 trees per acre before World War II to 500 to 2000 trees per acre for many species. With all commercial orchards, the mass-planting of singular varieties (with the exception of rows of pollenizer trees) has become the norm for management efficiency, resulting in vast monoculture orchards of the most commercially valuable varieties.

As a result of the trends in orchard history since World War II, earlier orchards with widely spaced trees on seedling rootstocks now represent archaic horticulture. Thousands of fruit varieties have been lost as a result of the decreased number of varieties grown. These changes distinguish older orchards and fruit trees in the California state park system, as they represent earlier periods in the history of American horticulture. Historic orchards in California state parks are the repositories of rare varieties or strains of varieties, and are becoming rare examples of old fruit tree forms and layouts. The more orchards change, the more distinguished California or chards dating prior to World War II will become, and the more unique the experience of an orchard landscape of the 1940s or earlier will be for park visitors.

Figure 11: Historic almond orchard in Sutter Buttes State Park (CPDR, 2010).

Just as California orchards have been transformed since the 1700s, new orchards will continue to evolve in design. As we preserve orchards that are 50 years of age or older with significance and integrity, their cultural resource value will continue to grow in importance. Genetic biodiversity conservation combines with visitor education as the potential societal benefits. However, these resources are in decline, and the need for stabilization intervention is growing urgent.

Orchards and Fruit Trees as Managed Cultural Resources

Efforts to stabilize the condition of potentially historic orchards and fruit trees are supported by CDPR Cultural Resource Management (CRM) policy. Orchards, groups of fruit trees and single fruit trees eligible or potentially eligible for listing in the National Register of Historic Places are cultural resources. Orchards and fruit trees may be independently eligible as historic districts or sites, or as contributing features. CDPR CRM policy supports a process for management involving research, analysis and evaluation, leading to identification, stabilization, treatment, and preservation maintenance of cultural resources. CDPR stabilization, treatment and preservation maintenance philosophies for historic orchards and fruit trees are based on the *Secretary of the Interior's Standards for the Treatment of Historic Properties and Guidelines for the Treatment of Cultural Landscapes* (NPS, 1992).

The process of identifying historic orchards and fruit trees involves historical research, field work, and documentation, followed by analysis and evaluation of significance and integrity. The NPS historic context document *A Fruitful Legacy* provides a basis for analysis and evaluation of significance and integrity. The products of initial identification may be a cultural resources inventory or a consensus determination with the California State Historic Preservation Office (SHPO). A SHPO consensus determination may lead to the preparation of a

National Register nomination and subsequent listing through the Keeper in the National Register of Historic Places.

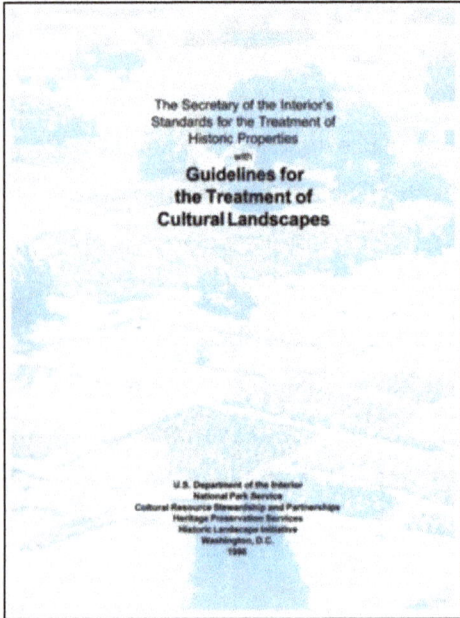

Figure 12: The *Guidelines for the Treatment of Cultural Landscapes* (NPS, 1992).

Beyond initial identification, CDPR policy calls for the protection and preservation of cultural resources using the techniques of preservation maintenance, repair and replacement in-kind, to perpetuate the same design, scale, form, and materials over time. Preservation guidelines are derived from the *Guidelines for the Treatment of Cultural Landscapes*. Under optimal conditions, preservation maintenance objectives for a specific orchard or fruit trees are outlined in an Orchard Management Plan. This type of plan describes the history, significance, and existing conditions of an orchard or fruit trees, defines management objectives and describes maintenance regimens for preservation, or how to implement restoration or rehabilitation treatments. Preservation maintenance actions for a historic orchard or fruit trees may involve all or some of the following activities: winter and/or summer pruning, weeding, aerating, mowing, cultivating, mulching, integrated pest management, fruit thinning, fruit harvesting, irrigating, fertilizing, monitoring, propagating, replanting, staking and record-keeping. Preservation maintenance activities are performed by qualified personnel with training in the cultural resource values of the orchard or fruit trees and the management objectives.

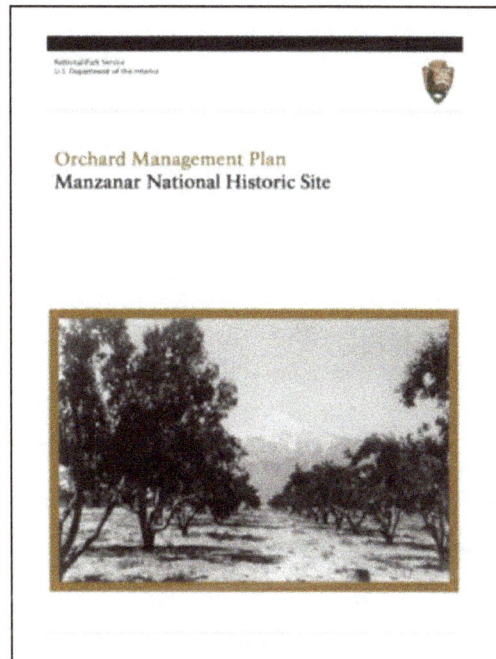

Figure 13: Example of an orchard management plan, a NPS document for Manzanar National Historic Site (NPS, 2010).

When a known or potentially historic orchard, a group of fruit trees, or a single fruit tree is in a poor or unstable condition, CRM policy calls for interim stabilization actions to be taken until the resource can receive preservation maintenance or ultimate treatment. Stabilization may also be undertaken before an orchard or fruit trees are formally evaluated as potential cultural resources. Under optimal conditions, stabilization is an interim step involving actions of temporary longevity to prevent the further deterioration of the resource. In less than optimal conditions, stabilization actions are used for a much longer duration, to hold onto potential resources until preservation prospects improve. Stabilization actions in historic orchards or fruit trees are more limited than the scope of preservation maintenance activities, and these include deadwood removal, bracing, sucker removal, encroaching vegetation removal, brush-hogging, mowing or aerating the orchard floor, irrigating and mulching. Stabilization actions are not supported by long term planning, and therefore do not include the propagation and planting of replacement trees, or quality fruit production. Germplasm conservation may be performed within the scope of stabilization. Field review by a qualified preservation specialist should be performed before stabilization actions are undertaken.

When the cultural resource management objectives of an orchard or fruit trees involve treatments other than preservation maintenance, a treatment plan is prepared. Four types of treatment are defined by CDPR CRM policy and the *Secretary of the Interior's Standards*: preservation, restoration, rehabilitation, and reconstruction. A treatment plan defines the type of treatment and provides recommendations or actions needed to implement the treatment. The treatment plan for an orchard or fruit trees may be part of a Cultural Landscape Report (CLR), which describes the history, significance, and integrity of a cultural landscape (Part I) and provides a treatment plan for the landscape (Part II). The treatment plan for an orchard or fruit trees that are part of a larger cultural landscape should be consistent and compatible with the treatment, significance, and period of significance of the landscape as a whole. An Orchard Management Plan may be prepared after a CLR to provide more detailed information on the implementation of a treatment plan.

The *Guidelines for the Treatment of Cultural Landscapes* provide standards for the four types of treatment.

- *Preservation* standards require retention of the greatest amount of historic fabric, including the landscape's historic form, features, and details as they have evolved over time.

- *Restoration* standards allow for the depiction of a landscape at a particular time in its history by preserving materials from the period of significance and removing materials from other periods.

- *Rehabilitation* standards acknowledge the need to alter or add to a cultural landscape to accommodate continuing or new uses while retaining the landscape's historic character.

- *Reconstruction* standards establish a framework for recreating a vanished or non-surviving landscape with new materials, primarily for interpretive purposes.

The treatment plan for an orchard or fruit trees should be based on clearly defined management objectives that are compatible with the type and level of significance of the property. Examples of management objectives include identifying a goal of partially restoring an orchard while protecting an adjacent archaeological site, or having a cooperating association produce and distribute fruit while preserving the historic character of the orchard. Objectives may include the preservation of an orchard landscape but excluding fruit production (through de-blossoming or immature fruit removal) to prevent wildlife interactions. The treatment plan may include drawings and construction details for the removal and replacement of fruit trees or encroaching vegetation, or for the re-establishment of orchard floor vegetation, circulation routes, boundary demarcations or other features.

Treatment plans should be prepared and implemented by personnel who meet the professional qualifications required by the *Secretary of the Interior's Standards*. Restoration and rehabilitation treatment actions for orchards and fruit trees go beyond the actions of stabilization, as they may involve the propagation of replacement trees using cuttings of appropriate scion wood and potential grafting with appropriate rootstock material. Restoration or rehabilitation treatments may involve replanting with custom-propagated replacement trees and the follow-up care of browse and sunscald protection, and irrigation until establishment.

SCOPE OF STABILIZATION

Stabilization Defined

In cultural resources management, the scope of stabilization is more limited than treatment. Stabilization is intended to be a short-term, interim endeavor until long-term treatment can be planned and implemented. Treatment actions include the full complement of those needed to reach and perpetuate the ultimate goals for preservation in perpetuity. Stabilization is limited to the complement of actions performed to prevent the further deterioration in condition of cultural resources.

Stabilization is performed when orchards or fruit trees are known to be, or are suspected of being potential cultural resources, and their condition has deteriorated from good or stable, to fair or poor. Stabilization may be performed before historical research and analysis and evaluation have been completed, in order to prevent the loss of potential resources.

Stabilization actions are, by definition, actions that do not alter the integrity of the potential or known cultural resource, or cause any loss of information, while at the same time, arrest deterioration in condition.

Figure 14: A historic pear orchard before stabilization at Manzanar National Historic Site. Note presence of deadwood, root suckers, and overgrown orchard floor (NPS, 2005).

Figure 15: Five years later, the orchard has been stabilized. Note the absence of deadwood, root suckers, and clean orchard floor (NPS, 2010).

Health Stressors

The concept of *removing health stressors* is a key to stabilizing the condition of potential biotic cultural resources, such as fruit trees and orchards. Stressors are factors that cause unfavorable living conditions over sustained periods, and undermine a fruit tree's natural ability to live, grow, reproduce or heal from predation, infection or mechanical damage. In the absence of underlying stressors, fruit trees have physiological strategies to adapt to the challenges of growth, reproduction and repair. In the presence of stressors, fruit trees have diminished resilience and their deterioration in condition is much more difficult to reverse. The old fruit trees in California state parks are part of the history of horticulture that has for several thousands of years adapted trees to respond to cultivated conditions. Fruit trees have lost some of the wild characteristics of resilience in order to bear more abundant and tasty fruit. As a result, fruit trees require more cultivated conditions than native or naturalized plant species. Cultivated conditions are those with less competition from other vegetation and adequate soil fertility. The process of eliminating stressors restores the more favorable, if not ideal cultivated conditions, in which fruit trees have been designed to live.

The most common health stressors are:

1) Encroaching Competitive Vegetation

2) Structural Instability of Fruit Trees

3) Reservoirs of Disease Infection on the Site

The presence of any or all of these stressors will reduce the resilience and vitality of fruit trees and limit their ability to respond to the stabilization actions of pruning, fertilizing and watering. The removal of underlying health stressors must be addressed in order to arrest deterioration, and have the trees respond to horticultural practices. Generally, further deterioration in the condition of fruit trees can be arrested by removing stressors alone, and performing no further stabilization actions. Fruit trees in poor condition can be stabilized in poor condition, and trees in fair condition can be stabilized in fair condition, by removing health stressors and therefore preventing further deterioration or loss. For improvement in condition from poor to fair, or from fair to good, additional stabilization actions such as pruning and irrigating must be employed. The most common health stressors of old fruit trees are outlined below.

Encroaching Competitive Vegetation

Encroaching competitive vegetation is any vegetation growing within the root or canopy zone of the fruit tree that is consuming water, nutrients, light or the physical space needed by the fruit tree, or sheltering pest organisms. Orchard fruit trees have been designed by horticultural selection to require their own physical space devoid of competitive vegetation, with the exception of compatible orchard floor ground cover plants. Encroaching competitive vegetation may include brushy undergrowth, overhanging overgrowth or root suckers and water sprouts from the old fruit tree itself. Removing the health stressor of encroaching competitive vegetation is generally one of the first needs in most unmaintained old orchards, and reducing or eliminating competitive vegetation is an important first step in arresting deterioration and rendering the trees responsive to further stabilization actions.

Figure 16: Historic pear orchard at Jack London State Historic Park is threatened by encroaching vegetation, creating unstable conditions (NPS, 2007).

Structural Instability of Fruit Trees

Structural instability is due to major defects in the main scaffold limbs, trunk or roots of the fruit tree, leading to threat of leaning, partial uprooting or complete collapse. Defects such as loss of bark, cavities in limbs, hollow trunks, detachment from the roots, or root damage can all result in structural instability. The presence of a hollow trunk is not a life-threatening condition, as the tree's conductive tissues for water and nutrients are located just beneath the bark.

Figure 17: A structurally unstable historic pear tree was toppled by wind-throw. Propping or bracing could have saved the tree (NPS, 2005).

Structurally unstable fruit trees are extremely vulnerable to snow and ice storms, windstorms or wildlife toppling, however, and are commonly killed by collapse. Structurally unstable trees are stabilized by propping or cabling individual limbs or bracing the whole tree to prevent leaning. Cabled, propped or braced trees can achieve a healthy condition, and this is a critical first step in protecting structurally unstable trees.

Reservoirs of Disease Infection on the Site

Reservoirs of disease infection are the sinks of life-threatening pathogens within the orchard or fruit tree site that have accumulated in the absence of maintenance. Life-threatening pathogens are those that affect the tree's vital physiological processes, and are not limited to just attacking fruit. Sinks of infection include deadwood attached or hanging in fruit tree canopies, downed dead trees and accumulated debris, or severely infected living trees within or adjacent to the orchard or fruit trees. Removal of these reservoirs of infection is critical to alleviate the threat of mortality, and to promote the natural longevity of the tree.

Natural Longevity of Fruit Trees

Orchard fruit trees have a natural longevity that is dependent upon species and the type of root system. Environmental factors such as health stressors and lack of maintenance combine with natural longevity to influence the lifespan. The following figure illustrates the spectrum of natural variation in longevity with species, varying from more than 200 years with olive trees to generally less than 50 years with peach trees. Note that the species commonly found in California state parks include apple, pear, orange and fig with the potential for a 150 to 200-year lifespan and apricot, cherry and plum with the potential for a 100 year lifespan.

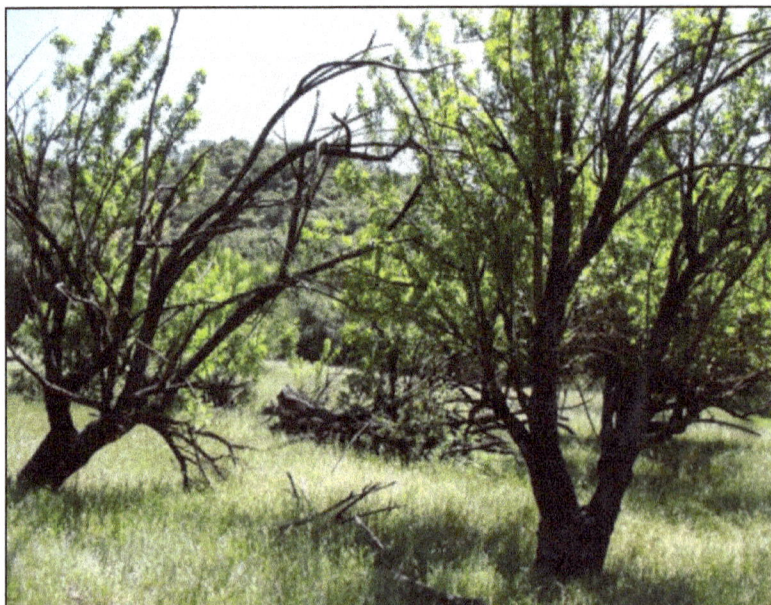

Figure 18: Presence of deadwood and downed debris are sources of infection among these historic almond trees at Sutter Buttes State Park (CDPR, 2010).

The influence of the root system type is for greater lifespan when fruit trees are grown on their own roots rather than a rootstock. For grafted trees, those grafted to seedling rootstocks are more long-lived than trees on dwarfing rootstocks. It should be noted however, that the quality of the graft union can influence longevity, even with seedling rootstocks. Poor unions can invite disease and cause premature death of the tree. As the predominant type of fruit trees before World War II were those grown on seedling rootstocks, most old fruit trees in California state parks have greater natural longevity than their contemporary counterparts on dwarfing rootstocks. Dwarfing rootstocks reduce the vegetative vigor of the scion or aerial part of the fruit tree, causing the tree to reproduce and yield fruit sooner in its life, but also shortening the lifespan of the tree. Un-grafted fruit trees grown on their own roots have the greatest longevity of all types, though few orchard species have been grown this way. Olive, fig and peach are among the few species that can be grown on their own roots, and for olive and fig trees, the upper limit of their longevity is so great as to be unknown. Self-rooted, ancient olive trees surviving in Croatia, Greece, Italy and Portugal have dendrochronology dates of 2000 to 3000 years old. A fig tree in Sri Lanka, the Jaya Sri Maha Bodhi tree, is the oldest living human-planted tree in the world with a documented planting date in 288 BCE, or 2,300 years old.

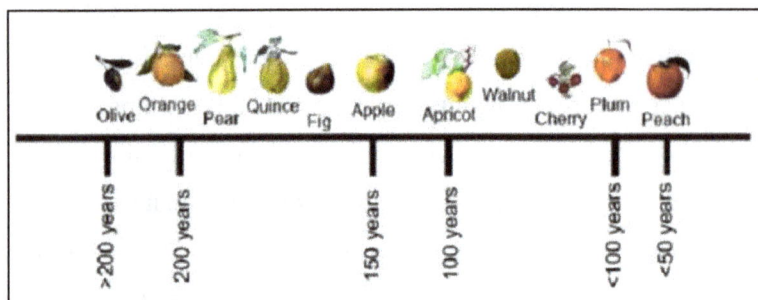

Figure 19: Diagram showing the spectrum of natural longevity in fruit trees is species-dependent, but is also affected by type of root system (NPS, 2008).

Condition Assessment

The stabilization of old fruit trees and orchards begins with the process of condition assessment. The condition assessment is an essential first step, as it provides managers with baseline information about the total number of living and dead trees, tree identity, location, general condition class and the various threats and risk of losses to potential resources. The condition assessment process is not designed to answer questions about historic significance and integrity, however, but to provide managers with the information needed to establish priorities for stabilization and to begin to perform and track stabilization actions. The following section provides guidance on how to perform a condition assessment of old fruit trees and orchards.

The goal of a condition assessment is to assign a condition class of "good", "fair", "poor", or "dead" to each fruit tree, as a result of observing and documenting the field conditions and creating a unique identity for each tree. The recommended protocol for assessing condition guides the assessor to "look beyond the canopy", and examine the orchard floor in the vicinity of the tree, as well as the roots, trunk, canopy and area above the canopy, in order to identify deficiencies and summarize the general status of each tree with a condition class. The condition classes of old fruit trees and orchards, like other cultural resources, are based on a measure of relative stability or instability, rather than industry standards for the acceptable condition of new trees or nursery stock. The condition classes are defined as follows:

Fruit Tree Condition Classes

Good: The tree has new growth at the terminal ends of shoots and only minor physical damage, defects, disease or insect damage, and/or only minor dieback or deadwood present.

Fair: The tree has decreased new growth with moderate physical damage, defects, disease or insect damage, or moderate dieback or deadwood present.

Poor: The tree is in a general state of decline with little or no new growth, major physical damage, defects, disease or insect damage, or major dieback or deadwood present.

The "Fruit Tree Condition Assessment Field Form" is the recommended dataset for recording each tree and guides examination in various zones in order to assess condition. The preferred format for dataset capture is electronic rather than hard-copy, due to the flexibility of data transfer to other management tools such as electronic databases and GIS. The field form dataset can be captured with the data-logger of a GPS unit, or directly input into the spreadsheet of an electronic tablet or other portable electronic device. If data are recorded in hard copy on the field form, later transfer into an electronic medium is recommended. The following text references the dataset of the field form.

Figure 20: NPS and CDPR staff use the fruit tree condition assessment field form to determine this historic pear tree is in fair condition (NPS 2007).

FRUIT TREE CONDITION ASSESSMENT FIELD FORM

Park: _____ Date: _____

Location: _____ Inspected by: _____

Species _____ Variety _____

Field ID #: _____ Diam.: _____

Tree significance (check): | Individual | Contributing feature | Non-contributing feature | Unknown |

Condition Assessment:

_____ Good: Good growth with minor physical damage, defects, disease or insect damage, or minor dieback/deadwood

_____ Fair: Decreased growth with moderate physical damage, defects, disease or insect damage, or moderate dieback/deadwood

_____ Poor: General state of decline with little or no growth, major physical damage, defects, disease or insect damage, or major dieback /deadwood

_____ Dead: Greater than 90% of crown dieback with no growth

Insert Photo

Inspection:

Zone	Description	Inspection Factors			
Zone 0: Orchard Floor	Ground beyond dripline	overgrown groundcover	rodent holes	gopher mounds	grade disturbances
		encroaching vegetation	accumulated debris	drainage issues	
Zone 1: Root System	Ground within dripline	root damage	accumulated debris	loss of soil	root suckers
		early fruit drop	exposed roots		
Zone 2: Trunk Base	Intersection of roots with trunk	loss of bark	cavities	fruiting bodies	cracks or splits
		girdling	suckers	root suckers	wildlife damage
		soil accumulation	trunk flare buried		
Zone 3: Main Trunk	Trunk up to scaffold limbs	unbalanced scaffolds	moss/lichen cover	pack rat nests	decay or cavities
		leaning trunk	deadwood	water sprouts	loss of limbs
Zone 4: Canopy	Scaffold limbs, branches and foliage	deadwood	pests	foliage discolored	foliage curled
		% live canopy	diseases	early leaf drop	foliage sparse
		unbalanced canopy	dieback of terminal shoots		
Zone 5: Above Canopy	Area above canopy	encroaching vegetation	over shading		

Recommendations:

_____ Prune/remove deadwood: _____

_____ Install support system/cable/brace: _____

_____ Monitor for change: _____

_____ Potential future removal: _____

Figure 21: Recommended fruit tree condition assessment field form (NPS, 2009).

Fruit Tree Condition Assessment Field Form

o   Identification Data

The most critical information in the field form dataset is the "Field Identification Number", the "Location" and the "Date" of assessment. These data initiate the means for future tracking of the tree.

The "Field Identification Number" is a unique identifier assigned to each tree. The identifier may be composed of letters and numbers, potentially incorporating the site location, row or column location (within an orchard grid) as well as tree number. The Field Identification Number may be attached to each tree with an easily visible metal plant tag, however, the tag should not be relied upon as the only means of later reconnaissance (i.e. a digital site map with field identification numbers is a more reliable source for future reconnaissance).

The "Location" data should be captured with at least the minimum level of accuracy to allow for future reconnaissance. Ideally, location data has sufficient accuracy to allow for projection on a map, by capturing geographic coordinates such as UTMs or latitude/longitude data.

The "Date" data is the date on which the tree is examined. This data field is updated upon future condition assessments.

The "Park" data is the name of the park or unit in which the tree is located.

The "Inspected By" data is the name(s) of the assessor(s). While different assessors may capture the data, consistency in approach between assessors is very important, to avoid disparities in the assignment of condition classes.

The "Species" data is the basic taxonomic rank of the tree, such as almond, apple, apricot, fig, lemon, olive, orange, pear, peach, persimmon, pomegranate, and walnut, etc. While fruits and leaves are the most immediate indicators of species, tree size, form and bark are also diagnostic for trained horticulturists. The Species data field should be left blank until a positive identification can be made by a trained individual.

The "Variety" data is the taxonomic rank below species, such as "Red Delicious", "Bartlett", "Eureka" or "Navel". Variety is usually determined through examination of fruits, and is generally not identifiable in the absence of fruit. In addition, old fruit trees with low vigor may produce poorly developed fruits that are uncharacteristic of the variety. The Variety data field should be left blank until a positive identification can be made by a trained individual.

The "Diameter" data is the diameter of the trunk at the location just below the point of attachment of the lowest scaffold limbs, rather than at breast height, which is the more common arborist's standard for tree diameter. Fruit trees often have short trunks, with the scaffold limbs borne within 3 feet from the ground and therefore the concept of trunk diameter at breast height is probably irrelevant. The diameter can be measured approximately using a tape measure sighted perpendicularly to the trunk, or by measuring the circumference and converting into the diameter by dividing the circumference by 3.14.

The "Tree Significance" data is a check mark indicating whether the tree is known to be historically significant as an individual, or as a contributing feature to a larger significant property. Old fruit trees that have lost integrity or newer fruit trees that have been planted or naturalized on a site may also be documented in the condition assessment process, and so "non-contributing feature" for known non-historic trees, may be checked. The condition assessment process informs but does not directly determine historic significance and integrity. A determination of significance and integrity is made as a result of historical research and the analysis and evaluation of existing conditions with respect to historic conditions. Significance may be "unknown", and this data field should be checked until the full process of determination is performed.

o   Condition Assessment Data

The condition class of the "Condition Assessment" dataset is assigned as a result of completing the inspection, i.e., it is the last dataset to be completed. The class is assigned based on the aggregate of the presence or absence of new growth and the combined deficiencies of all tree growth zones. The classes represent a spectrum of stability, and are relative assessments. The condition class is used in setting priorities for stabilization actions of individual trees, and therefore it is important to be consistent in applying the definitions of class from tree to tree. A photo that illustrates the condition of the tree is useful for tracking changes in condition in the future.

o   Inspection Data

The "Inspection" dataset guides the assessor to examine the fruit tree within different growth zones, and to record the presence of specific deficiencies in these zones.

"Zone 0" corresponds to the orchard floor within the vicinity of the tree but outside the root zone or dripline of the tree canopy. In this zone, the assessor is looking for deficiencies outside the healthy growing conditions of an open, low ground cover, (or possibly no ground cover in some orchards) with a continuous, stable substrate.

Figure 22: Condition assessment field form "Zone 0" is the orchard floor immediately outside the tree canopy. In the North Coast Redwoods District, Zone 0 of this apple tree is unstable, due to colonizing blackberries (CDPR, 2010).

Deficiencies found in this zone include overgrown ground cover (taller than six inches) or encroaching vegetation that is competitive with the fruit trees, or evidence of deficient substrate conditions, such as wildlife colonization (burrows, nests) or drainage problems (puddles, gullies).

"Zone 1" corresponds to the root system of the individual fruit tree, found within or just beyond the edge of the dripline. In this zone, the assessor is looking for deficiencies outside the healthy growing conditions of an open, low ground cover (or possibly no ground cover) with a continuous, stable substrate covering the tree roots. Deficiencies found in this zone include exposed, damaged roots, accumulated debris (either vegetative or non-vegetative), root suckers or unseasonably early fruit drop, indicating the presence of health stressors.

Figure 23: Condition assessment field form "Zone 1" is the orchard floor within the dripline of the canopy. At Manzanar National Historic Site, Zone 1 of this pear tree is stable with no encroaching vegetation. A drip irrigation line can be seen around the dripline of the canopy (NPS, 2010).

"Zone 2" corresponds to the base of the tree trunk, where the trunk meets the roots. In this zone, the assessor is looking for deficiencies outside the healthy condition of a sound trunk flare meeting the substrate. Deficiencies in this zone include the absence of a trunk flare when the base of the trunk is buried by accumulated substrate; damage to the trunk due to wildlife browsing, with girdling or loss of bark; signs of pests and diseases such as cankers or fruiting bodies; signs of structural instability such as cracks, splits or cavities; and/or signs of encroaching vegetation such as root suckers or competitive undergrowth.

Figure 24: Condition assessment field form "Zone 2" is the base of the trunk at the point of attachment to the roots. In Manzanar National Historic Site, Zone 2 of this pear tree is unstable, due to the presence of a rodent burrow and root suckers (NPS, 2005).

"Zone 3" corresponds to the trunk of the tree between the base and the point of attachment of the scaffold limbs. In this zone, the assessor is looking for deficiencies outside the healthy condition of a sound, upright trunk with continuous bark cover. Deficiencies found in this zone include signs of structural instability due to a leaning or hollow trunk or with cracks and splits; signs of pests and diseases such as cankers, fruiting bodies or wildlife nests; signs of unmanaged vigor such as epicormic growth (shoots arising directly from the trunk, rather than the terminal branches); or signs of dieback such as loss of scaffold limbs, or hanging or attached deadwood.

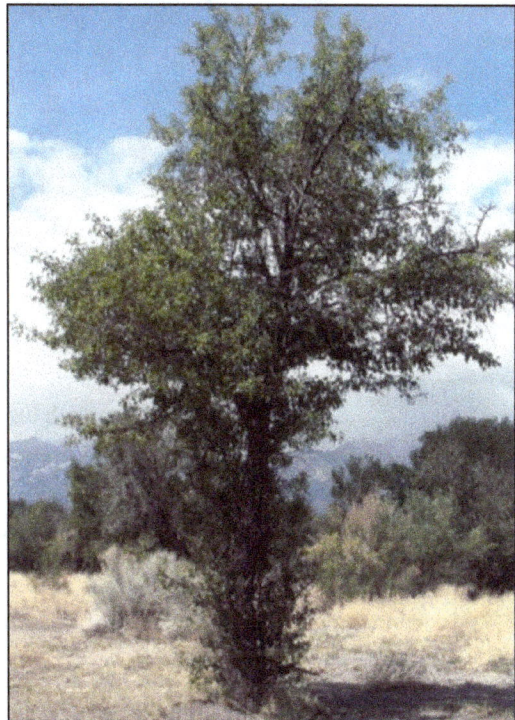

Figure 25: Condition assessment field form "Zone 3" is the tree trunk. In Manzanar National Historic Site, Zone 3 of this pear tree is unstable, due to the presence of multiple shoots (epicormic growth) on the trunk, and root suckers (NPS, 2005).

"Zone 4" is a large zone, corresponding to the major scaffold limbs, branches, shoots and foliage of the tree. This is the zone that usually captures our attention. In this zone, the assessor is looking for deficiencies outside the healthy condition of sound, structurally balanced scaffold limbs, with a well-branched (also known as a "well-feathered") canopy, with a majority of live branches and shoots, well-covered with live foliage.

Figure 26: Condition assessment field form "Zone 4" is the canopy of the tree. At this pear tree in Jack London State Historic Park, Zone 4 is unstable due to severe dieback, attached deadwood, and sparse foliage (NPS, 2008).

Deficiencies in this zone include signs of dieback such as deadwood, loss of limbs, sparse foliage, or no new or diminished new growth (look for the length of new wood at the shoot tips and compare to the previous year's growth; six inches is a measure of "good" growth for an old fruit tree); signs of pests and diseases such as cankers, fruiting bodies or wildlife nests; signs of competitive vegetation such as encroaching plants or water sprouts (vertical, unbranched vegetative shoots that compete with the desired canopy of the tree); or signs of poor fertility or drought such as discolored, curled or sparse foliage, premature leaf drop or dieback of terminal shoots.

"Zone 5" is the area above or around the tree canopy which provides light and air and a physical space for new growth. In this zone, the assessor is looking for deficiencies outside the healthy condition of a complete absence of other vegetation, objects or features that may block light and air or impinge upon the space for the tree to grow. Deficiencies in this zone include encroaching competitive vegetation, both direct (adjacent) and indirect (not adjacent but with the ability to cast more than six hours of shade per day), or encroaching objects or features such as temporary structures, stored or stockpiled materials and equipment.

Figure 27: Condition assessment field form "Zone 5" is the area above or around the tree canopy. At this plum tree (lighter green canopy) at Jack London State Historic Park, Zone 5 is unstable, due to the over-shading presence of a Live oak tree (NPS, 2008).

o   Field Condition Assessment Recommendations

The "Recommendations" dataset prompts the assessor to record any immediate or particular needs of the individual tree for stabilization or the potential for future removal, in the case of a condition class of "dead".

Upon completion of the "Inspection" part of the dataset, the condition class is assigned, based on an aggregation of the deficiencies observed in all tree growth zones. After compiling the data for all trees in an electronic medium, the first round of condition assessments is complete.

Re-assessing Condition

The condition of biotic cultural resources such as old fruit trees and orchards is dynamic, and the condition assessment should be periodically repeated to maintain accurate management records. An annual condition assessment is recommended to note any changes in condition. A comprehensive condition assessment is recommended every five years, to include a review of the complete dataset in the field form. Each time the complete dataset should be updated, to reflect current conditions and priorities for stabilization.

Setting Priorities

The condition assessment data are used to set priorities for stabilization actions. The data provide the manager with the number, location and general condition classes of fruit trees on a site or within an orchard. Ideally, the condition assessment data for all fruit trees on a site or within an orchard are compiled electronically into one database, searchable by tree identification number, species, variety or condition, and tree locations are projected on a site map.

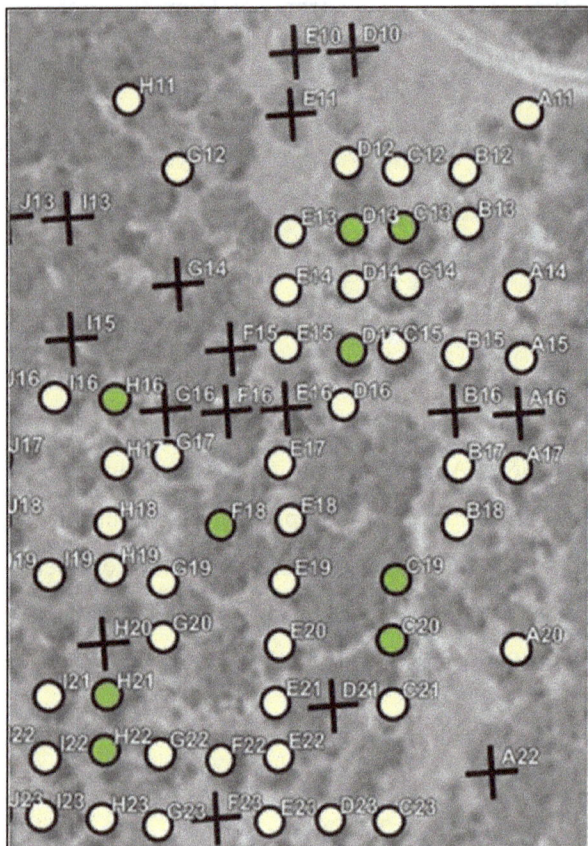

Figure 28: Sample segment of an orchard site map, showing tree identification number, location, and condition classes (green = fair, yellow = poor, cross = dead). In this orchard, all of trees are the same species and variety (NPS, 2009).

Analysis of the condition assessment data highlights common problems within zones, such as deficiencies throughout the orchard floor, or within the root zones. These data help the manager understand the extent and severity of threats, and strategize on an approach to arresting further deterioration. A recommended approach is to remove common health stressors first, as this will render all of the trees more responsive to subsequent stabilization actions. Stressors common in neglected fruit trees or orchards include encroaching competitive vegetation, structurally unstable trees, and reservoirs of pests and diseases. A big impact on stability can be made by removing encroaching vegetation by brushing and mowing, propping or bracing leaning trees, and removing downed debris and hanging deadwood. These actions are detailed in the "Stabilization Actions" section. With the exception of propping or bracing, these actions can be performed by non-specialists, such as a group of trained volunteers equipped with personal protective equipment and supervised by a qualified staff member.

Where a whole site or whole orchard cannot receive stabilization to remove common health stressors due to limited capacity or other resource protection issues, the strategy to remove stressors can be refined to prioritize geographic areas with the greatest concentration of fruit trees in good or fair condition first. The idea of protecting trees in best condition first may seem antithetical, but is based on the concept of triage. The idea is to prevent trees in fair condition from deteriorating to poor condition, and recognizes that trees in fair condition have a greater probable lifespan and potential for improvement to good condition than trees in poor condition. Exceptions to this approach include situations where trees in poor condition have a higher cultural resource value or are most accessible or visible for visitors' experience than trees in good or fair condition, and should therefore receive priority for stabilization.

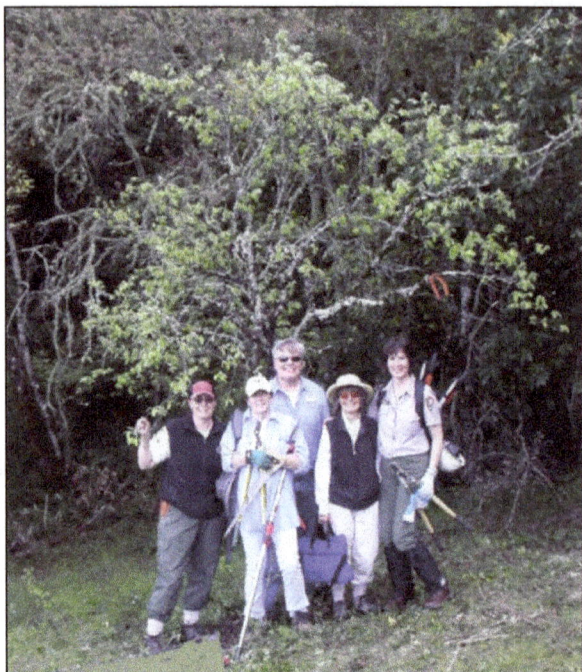

Figure 29: CDPR and NPS staff celebrate the successful stabilization of a historic quince tree at Jack London State Historic Park. Stabilization included encroaching vegetation and deadwood removal (CDPR, 2010).

After prioritizing the removal of common health stressors, individual fruit trees should receive stabilization actions in a recommended sequence of those in good and fair condition before those in poor condition or those that are dead. Trees that are standing dead are prioritized for removal if they are a potential hazard (i.e., have the potential to hit a valuable target), or if they are hosts to pests and diseases, but should first be adequately documented.

Figure 30a: One of two site maps of the historic orchard system at Jack London State Historic Park showing two approaches to stabilization. This map shows the area to be fully stabilized (NPS, 2009). The following map shows a partial stabilization alternative (NPS, 2009).

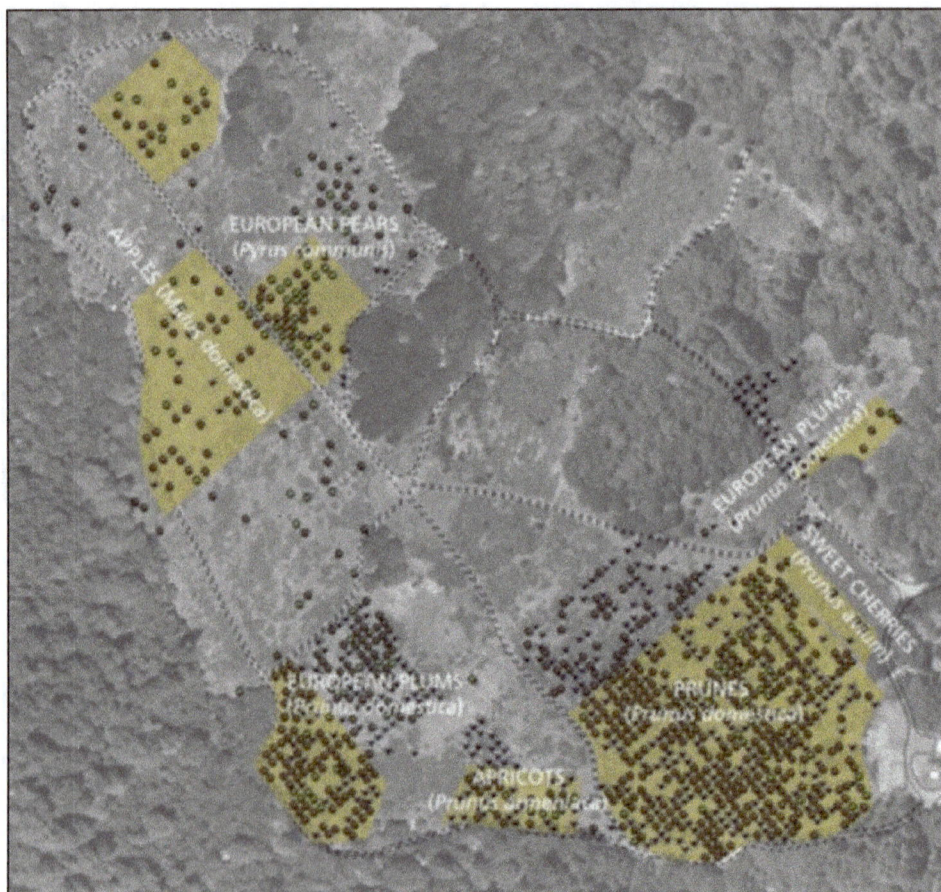

Figure 30b: One of two site maps of the historic orchard system at Jack London State Historic Park showing two approaches to stabilization. This map shows how a smaller scope: partial stabilization, could be strategized, prioritizing areas with trees in better condition (fair rather than poor), and most visible from visitor trails (NPS, 2009).

Managing Stable Conditions

The ability to manage stable conditions is dependent upon adequate information and planning for cyclic stabilization actions. While stabilization is intended to be an interim measure, the interim may extend for a period of years before treatment planning can be performed and implemented. During the interim, stabilization actions may need to be repeated in seasonal or annual intervals, depending on the condition. Planning for cyclic stabilization actions requires sustained record keeping built upon the original condition assessment dataset.

			California Department of Parks and Recreation Site Name Fruit Tree Stabilization Record							
Tree ID	Condition	Deficiency	Dead-wood Prune	Sucker removal	Brush Hog	Aeration	Fertile Mulch	Prop/Brace	Cavity repair	Other Action Taken
A1	Fair	Cracks/splits	01/01/2012	01/01/2012	02/01/2012	02/05/2012	02/06/2012			
A2	Poor	Decay	01/01/2012	01/01/2012	02/01/2012	02/05/2012	02/06/2012			
A3	Fair	Loose/cracked Bark	01/01/2012	01/01/2012	02/01/2012	02/05/2012	02/06/2012			
A4	Poor	Previous Failure	01/01/2012	01/01/2012	02/01/2012	02/05/2012	02/06/2012	01/15/2012		

Figure 31: Sample fruit tree stabilization record, tracking the date and type of stabilization actions by tree (NPS, 2011).

Changes in condition are tracked through annual or comprehensive condition assessments, and stabilization actions are recorded over time. The figure above illustrates an example of a fruit tree stabilization database record. A database can be converted into a GIS map when it contains geographic coordinates for each unique tree identification number. The goal in managing stable conditions is twofold: 1) to prevent further deterioration in condition, and 2) to improve condition from poor to fair or fair to good. Managers can use the following standards for stable fruit trees/orchards as a guide to achieving and sustaining stable conditions.

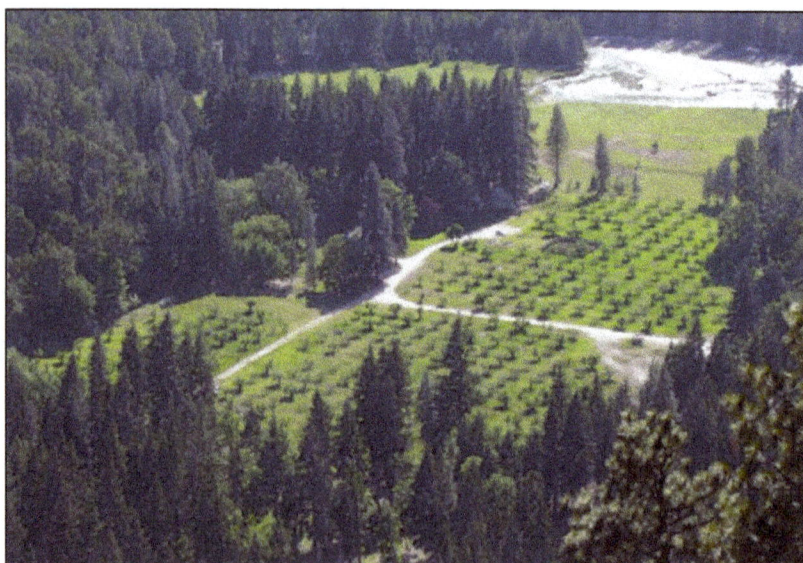

Figure 32: Aerial photo depicting stable historic orchard conditions at the Buckner Orchard in North Cascades National Park Service Complex, WA (NPS, 2009).

Standards for Stable Orchards and Fruit Trees

Stable orchards and fruit trees in good condition...

...have a clear, open site or orchard floor, with a low ground cover (or no ground cover in some orchards) with no accumulated debris or encroaching vegetation over six inches in height, as the ground cover is brushed or mowed.

...have no overhanging or encroaching overstory vegetation that casts shade for more than six hours per day during the growing season, as overshading vegetation has been headed back, thinned or removed.

...have no structurally unstable or leaning trees as all vulnerable trees have been cabled, propped or braced.

...have no reservoirs of pests and diseases, as all downed debris, hanging or attached deadwood, wildlife nests and burrows have been removed.

...have no symptoms of severe drought or severe nutrient deficiency, as water and nutrients are provided to deficient trees through irrigation and nutritional mulch.

...have no evidence of severe wildlife damage, as vulnerable trees are protected from wildlife browsing by physical barriers or chemical repellants.

Germplasm Conservation

Conservation of germplasm is recommended as part of the scope of stabilization. Germplasm conservation preserves the genes of each variety and each species (the full complement of genotypes) in the orchard in perpetuity. Conservation can be achieved by two means, one, through a living collection of trees representing all of the genotypes in the orchard and maintained off-site, such as in a plant nursery, and two, through cryogenic means, involving use of the national system of USDA National Plant Germplasm Repositories. Cryogenically conserved germplasm is plant tissue held at sub-zero temperatures in liquid nitrogen, so that it can be thawed at any time later and used to propagate replacement trees in perpetuity.

Germplasm conservation uses fruit tree cuttings from the scion. Cultivated fruit trees consist of two individuals grafted together: the scion, or aerial parts of the tree (trunk, limbs, canopy), and the rootstock, the root crown at the base of the trunk and the root system. The aerial parts of one tree and the roots of another are joined by grafting when each tree is approximately one year old. The grafted tree is then grown for one to two more years before planting in the orchard.

Most California state park fruit trees date from the period between the 1880s and the 1940s when fruit trees were designed to grow to a full size. These trees were grown on seedling rootstocks rather than "clonal dwarfing" rootstocks, which

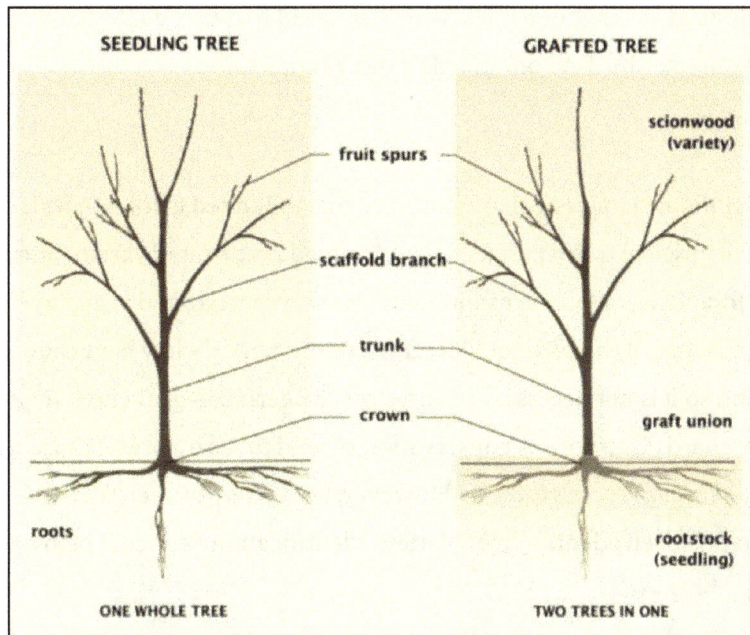

Figure 33: Diagram distinguishing a seedling fruit tree from a grafted fruit tree, and indicating the scion and rootstock components of a grafted tree (NPS, 2009).

were used in California for new orchard plantings after World War II. Seedling rootstocks are derived from trees grown from seed. These are genetically unique individuals and have no variety. Seedling rootstocks can be identified as individuals of the "straight" species, such as Domestic apple, *Malus domestica* or European pear, *Pyrus communis*. These rootstocks confer vigor on the scion, stimulating the tree to attain a full size. Seedling rootstocks are contrasted with clonal dwarfing rootstocks which are used to dwarf contemporary fruit trees from two-thirds to one-third of the standard size.

Scions are clones of an original tree that exhibited favorable fruit characteristics. The original tree was selected and named a cultivated variety or cultivar, such as "Jonathan" or "Gravenstein" and was then propagated vegetatively to clone the genotype of the original. All scions of the same variety have the same genotype, and so it is not necessary to conserve the germplasm of every scion in the orchard. Germplasm conservation should focus on preserving *each variety within each species present*. However, the variety of each fruit tree may be difficult to positively identify (see "Variety Identification Services" below).

Germplasm cuttings are taken from dormant shoots with several replicates (multiple individuals) of the same species and same variety, during the winter period. For example, germplasm for the apple variety "Jonathan" can be taken from several Jonathan trees. Each set of germplasm should be placed in a labeled, zippered plastic bag with damp tissue paper, and then refrigerated until packaging and express mailing to the USDA Germplasm Repository can occur. Contact the Germplasm Repositories at the following addresses:

o **Apple Germplasm Conservation**

Plant Genetic Resources Unit

USDA, Agricultural Research Service

630 West North Street

Geneva, New York 14456

o **Pear Germplasm Conservation**

USDA, Agricultural Research Service

33447 SE Peoria Road

Corvallis, Oregon 97333-2521

o **Apricot, Plum, Prune and Cherry Germplasm Conservation**

USDA, Agricultural Research Service

One Shields Avenue

Davis, California 95616

Conservation services can be provided at the USDA National Plant Germplasm Repositories through the development of a cooperative agreement between California State Parks and USDA NPGR.

Variety Identification Services

The variety of a tree may be difficult to identify, due to the absence of fruit or the bearing of uncharacteristic fruit, common in trees with poor health. Variety identification services can be procured through the USDA National Plant Germplasm Repositories or the California Rare Fruit Growers. Relatively new scientific developments in DNA mapping now enable the identification of fruit varieties from the DNA of leaf tissue, rather than from the macroscopic appearance of fruit. The process, called "DNA Fingerprinting" is now available for pear, apple and some stone fruits. Contact the respective Germplasm Repositories previously listed for more information about DNA Fingerprinting for variety identification. The process makes use of shoot cuttings with new leaf tissue, taken when new leaves are just emerging in spring.

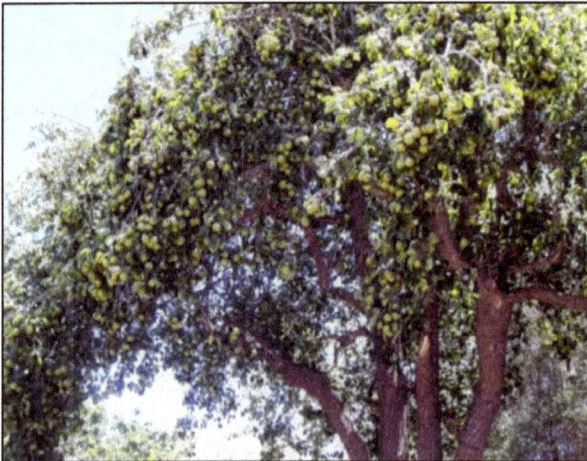

Figure 34: A historic pear tree at La Purisima Mission State Historic Park bears an abundance of fruit (CDPR, 2010).

The California Rare Fruit Growers may be able to provide variety identification services from fruit samples (collected and refrigerated in labeled zippered plastic bags and expressed mailed to the organization). Contact CRFG at the following address:

California Rare Fruit Growers, Inc.

The Fullerton Arboretum

California State University at Fullerton (CSUF)

P.O. Box 6850

Fullerton, CA 92834-6850

http://www.crfg.org

STABILIZATION ACTIONS

Encroaching Competitive Vegetation Removal

The general standard for a stable orchard floor is to have a continuous, herbaceous ground cover that is no taller than six inches. Herbaceous ground covers taller than six inches, with colonizing woody plants and/or root suckers, shelter wildlife and pest organisms and compete with fruit trees for light, water and nutrients. Another general standard is for the space around each tree canopy to be clear and unimpeded by overshading vegetation, such as the overstory canopy of adjacent trees. Directly adjacent vegetation or nearby trees that cast more than six hours of shade per day during the growing season, pose a threat to old fruit trees.

The goal of encroaching vegetation removal is to clear the orchard floor of any woody vegetation other than the fruit trees to be stabilized, and to reduce the height of the herbaceous ground cover to less than six inches, so that the orchard floor can be subsequently mowed. In addition, overshading canopies should be thinned or removed, and root suckers around the trunks or root zones of fruit trees should also be removed so that trunks are unobscured.

Figure 35: Stable orchard floor conditions in the historic Buckner Orchard in North Cascades National Park Service Complex, with low herbaceous ground cover and no encroaching woody plants (NPS, 2009).

Techniques for encroaching vegetation removal involve both power equipment and hand tools. However, regardless of the method used, vegetation removal should involve flush-cutting at the ground level rather than pulling or uprooting vegetation. As stabilization actions are limited to those that do not destroy information, ground disturbance is avoided to prevent the disruption or damage of potential archeological resources. Pulling woody vegetation also has the potential to disrupt the feeder roots of old fruit trees, and this should be avoided.

Consideration must also be given to the protection of natural resources. Ground-nesting birds, amphibians, reptiles, other wildlife and rare plants may have colonized the orchard floor, and their presence should be identified before clearing work begins. Collaboration with natural resource specialists is needed in scheduling vegetation removal to avoid or minimize impacts to natural resources.

For overgrown orchard floors containing tall woody plants such as Coyote brush or Blackberry, a brush hog is the most effective tool for low-cutting. Brush hogs can be rented or purchased in a range of sizes, varying from walk-behind self-propelled machines to larger attachments for ATVs or tractors. A walk-behind brush hog is most useful within the root zones of fruit trees, whereas a large, tractor-pulled hog should be limited to areas outside fruit tree root zones, to prevent damage to trunks or root zones.

Figure 36: Examples of a walk-behind brush hog (upper photo) and an ATV-pulled brush hog (lower photo), for removing woody brush on the orchard floor (DR Power Equipment, 2011).

Hand tools for vegetation removal include brush cutters, machetes and rakes. Brush cutters are like powerful loppers, and are effective for cutting woody vegetation at the ground level. Machetes are cleaver-like cutting tools, and rakes are used to aggregate and pile debris for removal from the site.

Root suckers emanating from the base of the tree trunk or within the root zone should be flush cut at the trunk or the ground level using pruners, loppers or a pruning saw (for very large suckers). Generally, pruners can cut through tissue less than a finger's width in diameter, and loppers should be used for larger material.

Figure 37: A common need in neglected orchards is root sucker removal from the base of trees (NPS, 2005).

Figure 38: Historic pear tree lower trunk at Manzanar National Historic Site after sucker removal. Note the suckers are flush-cut, and debris will be removed from the site (NPS, 2010).

Upon completion of vegetation clearing, the orchard floor may be mowed to refine the height of the vegetation, and increase access to the fruit trees. Vegetation clearing can be an enormous, daunting operation with the potential for resource impacts and physical challenges to remove the debris from the site. All of the debris should be removed to avoid accumulating reservoirs of pests and diseases, however, it may be processed into compost and returned to the site when cured. Ideally, vegetation clearing is only performed once during the stabilization period. Having to repeat the operation can be avoided by maintaining a low, herbaceous orchard floor through mowing. Mowing can be simplified by mulching within the driplines of fruit trees, and if repeated seasonally, will prevent re-colonization by woody plants.

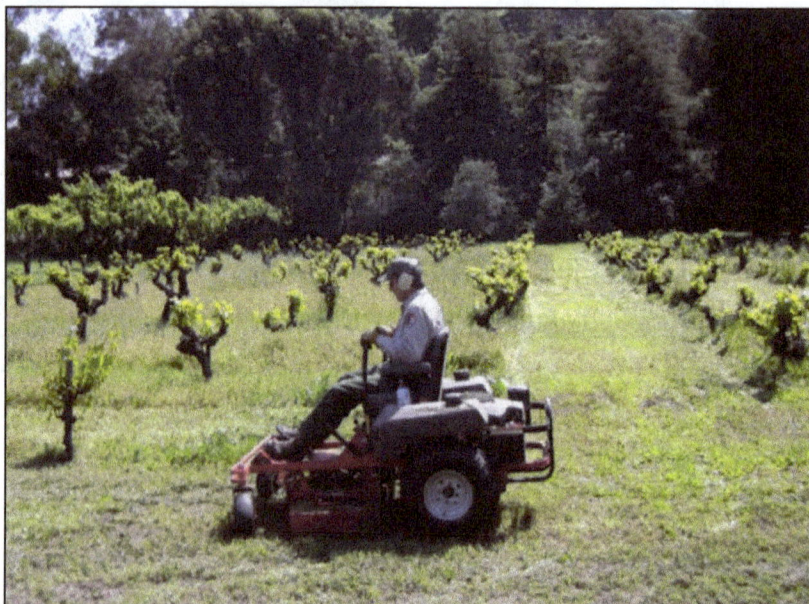

Figure 39: NPS staff member at John Muir National Historic Site demonstrates the use of a rider mower to mow within a low-headed grape vineyard, suitable also for orchard floors (NPS, 2006).

Deadwood Removal

A general standard for stable fruit trees is to have no deadwood within or around the trees, attached as dead limbs, branches or stubs, or hanging within the canopy. Dead and diseased wood adversely effects fruit trees by serving as reservoirs for pest and diseases, by consuming space for new growth, and interfering with the process of wound repair. The presence of deadwood imposes various burdens on old fruit trees that are alleviated when deadwood is removed. The goal of deadwood removal is to "clean" the tree, so that no deadwood is left behind. Dead and diseased wood, just like other debris, should be removed from the site to prevent the accumulation of pests and diseases. The following figure illustrates the goal of deadwood removal and the cleaning of a tree.

Dead and diseased wood removal is a pruning activity that can be performed at any time of year. Deadwood removal is possible using hand tools such as pruners, loppers and handsaws for smaller-diameter deadwood, as well as power tools for larger material. The hand-tool work may be a suitable activity for trained volunteers with personal protective equipment and a qualified supervisor. Tools should be sanitized with isopropyl alcohol before use and in-between contact with successive trees.

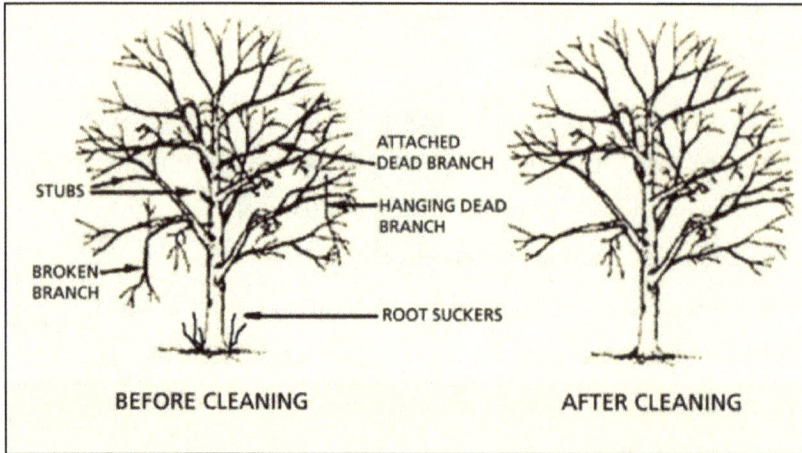

Figure 40: Diagram showing tree conditions before and after cleaning activities (NPS, 2011).

An important safety consideration is head protection when working beneath tree canopies with deadwood, or up a tree ladder. Large dead scaffold limbs and branches should be removed in sections only by the qualified operator of a chain saw, to avoid the physical disruption of hand sawing on an unstable tree, and reduce the safety hazard. A bucket truck or high-lift may be needed for the chain saw operator to reach high limbs.

Despite deadwood removal being possible at any time of year, it is a skilled activity that should be performed by adequately trained workers. Pruning cuts must be made in the correct location to allow healing to occur. Stubs that are left behind as deadwood will prevent wound repair and serve as gateways for disease infection. The following figures illustrate the correct location of pruning cuts to remove a dead scaffold limb, or a dead co-dominant scaffold limb. A co-dominant scaffold is a limb that arises from the same point as another scaffold limb, creating a weak joint. Larger deadwood, such as branched or scaffold limbs, should only be removed by a qualified chainsaw operator familiar with the following arborist's standards: ANSI A300 (Part 1) – 2001, "Tree, Shrub and Other Woody Plant Maintenance – Standard Practices (Pruning)". While smaller deadwood removal may be a suitable activity for volunteers with hand tools, the same standard applies.

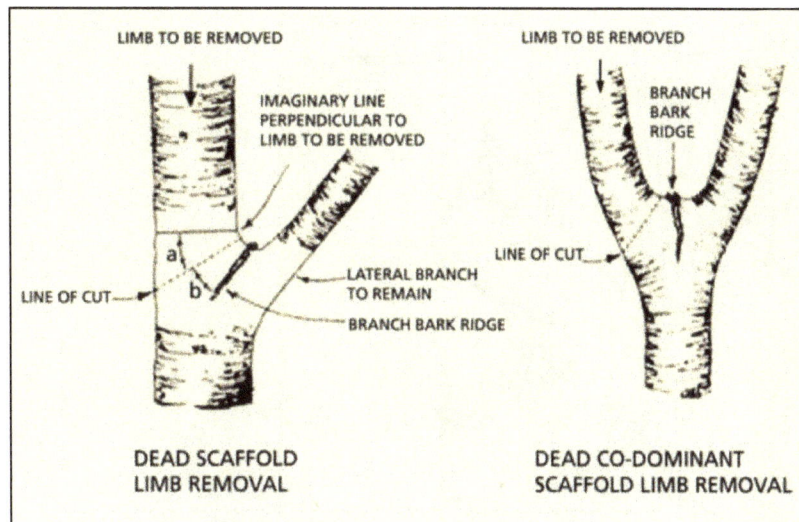

Figure 42: The recommended cut locations for the removal of a dead scaffold limb and a dead co-dominant scaffold limb. Note limbs over three inches in diameter should only be removed by a certified arborist (NPS, 2011).

A dead branch is removed by cutting immediately outside of the branch collar, leaving behind the branch protection zone from which new tissue will form to heal the wound. This is called the "Target Pruning Method" (see following figure). Inappropriate cuts that are too deep into the remaining live limb remove the branch protection zone and prevent subsequent healing. Inappropriate cuts too far outside the branch collar leave a stub of deadwood behind that invites decay. A dead scaffold limb is removed by making a diagonal cut at the limb/trunk union, determined by locating an imaginary line perpendicular to the limb to be removed. The diagonal cut is made approximately 45% down and away from the imaginary line, as shown in the figure.

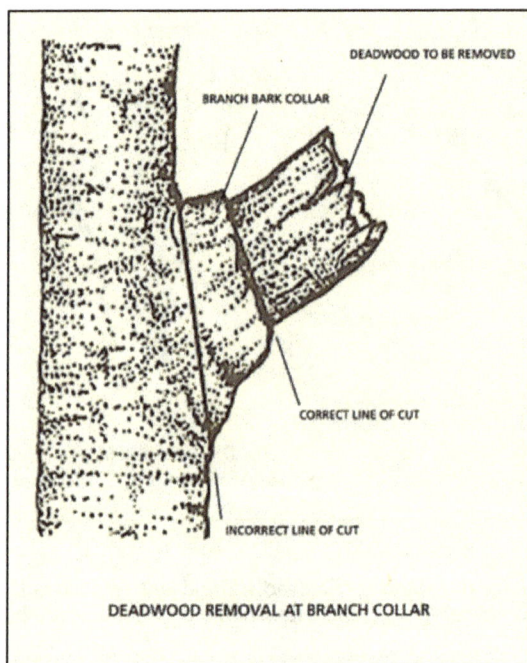

DEADWOOD TO BE REMOVED
BRANCH BARK COLLAR
CORRECT LINE OF CUT
INCORRECT LINE OF CUT

DEADWOOD REMOVAL AT BRANCH COLLAR

Figure 43: The Target Pruning Method for deadwood removal, where the cut is made immediately outside the branch bark collar (NPS, 2011).

Figure 44: Photo showing a poor pruning cut where a limb was removed by cutting inside the branch bark collar, i.e., too deep, preventing natural healing and causing subsequent infection (CDPR, 2010).

Cabling, Propping and Bracing Unstable Trees

The cabling, propping or bracing of structurally unstable trees are skilled activities that should be performed by qualified staff familiar with the following professional standard: ANSI A300 (Part 3) – 2001 – Tree, Shrub, and Other Woody Plant Maintenance (Support Systems; Cabling, Bracing and Guying).

Figure 45: A toppled historic pear tree at Fort Ross State Historic Park is still alive, but severely compromised in condition (CDPR, 2010).

Cables are installed within tree canopies to support unstable scaffold limbs. The cables may prevent a single limb from collapse, or prevent the crown (point of attachment of the scaffold limbs) from splitting apart. Cables are anchored directly to the tree at the trunk or to other scaffolds, or may be anchored to a vertical steel post.

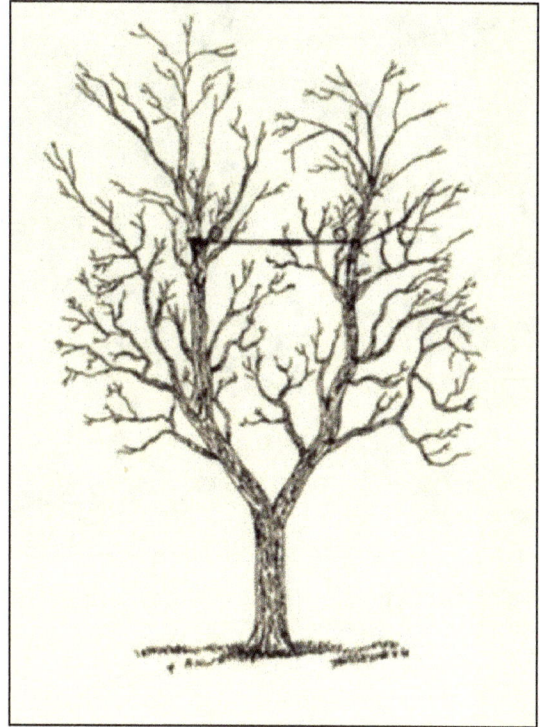

Figure 46: Diagram of cabling to prevent two scaffolds splitting apart (NPS, 2011).

Props are supports for individual limbs that are often temporary in nature, used until wound healing and new growth occurs. Prop materials include wood and steel posts and are generally held vertically in place by the weight of the limb. The point of contact of the prop with the limb is padded to prevent abrasion during movement.

Figure 47: Photo of a prop supporting a leaning apple tree, preventing the tree becoming severed at the roots (NPS, 2009).

Braces are supports for the trunk to prevent leaning, splitting or collapse. They include steel and wood posts with concrete footings, and guy wires anchored to footings. They also include steel bars inserted through trunks or scaffold limbs to prevent splitting. Braces can be permanent forms of support systems and are installed when old trees are no longer self-supporting or are vulnerable to collapse.

Figure 48: Examples of braces: vertical braces support a leaning tree (upper photo) and a horizontal brace prevents a tree from splitting apart (NPS, 2007).

Orchard Floor Management

Orchard floor management is important in sustaining stable conditions. Mowing, aerating, and mulching are actions that retain non-competitive, ground cover vegetation on the orchard floor and promote the penetration of air, water and nutrients to tree roots.

Mowing

Mowing is performed at regular intervals throughout the growing season, in order to retain a low ground cover and discourage the establishment of woody plants on the orchard floor. A low ground cover is less competitive with fruit trees for light, water and nutrients, and does not shelter potential pests as readily as taller vegetation. In addition, a low ground cover facilitates access to the trees for cleaning, pruning and harvesting activities, enabling both visitors and volunteers to enjoy the experience of the fruit trees. Mowing can be performed as frequently as once a week, if resources allow, or as little as once a month during the growing season in spring and early summer. Longer intervals allow for the re-establishment of woody plants that may resist cutting with a mower. Once woody vegetation has re-established on the orchard floor, a mower is typically inadequate and instead, a brush hog is needed to do the cutting.

The recommended equipment for mowing within a historic orchard is a riding mower with a compact size, low profile and easy maneuverability beneath tree canopies. A small tractor with a mower attachment is efficient for mowing between rows and columns of trees, but is generally too large to mow beneath fruit tree canopies.

Figure 49: An appropriately-scaled rider mower for maintaining the orchard floor in-between the rows and columns of trees in an orchard. Here, the operator is mowing a vineyard (NPS, 2006).

Canopies should not be limbed up to make space for the tractor. The fruit trees in California state parks that date from the 1880s to the 1940s have their canopies borne low by design, and this historic characteristic should be preserved. A riding mower has the greatest versatility for mowing between and within the rows and columns of trees. These mowers have height adjustments and can be raised sufficiently to deal with brushy stubble, low woody stumps or slightly uneven ground. In addition, some models can be fitted with a mulching attachment, to both mow and macerate the cuttings at the same time. Any tree debris such as pruning clippings should be removed before mowing and macerating, to prevent the spread of tree diseases. The need to mow beneath the tree canopies can be avoided by maintaining a layer of nutritional mulch within the dripline of the tree. Mulching is discussed below.

Aerating

Aerating is needed as a stabilization action where the site or orchard floor is highly compacted as a result of years of use by people and equipment, or the effect of weathering by surface water. Compacted soils appear and perform like hard pan, and resist the penetration of air and water to tree roots. Compact soils devoid of air spaces and water holding capacity support less root development and have less organic matter and microbial activity to release nutrients. The result of soil compaction is diminished fruit tree root and canopy growth, and water and nutrient deficiencies. Aerating should be performed after brush clearing and low-mowing, and preferably when the soil is damp but not saturated. Aerating should also be timed to occur before mulching with nutritional mulch.

Aeration is targeted primarily in the vicinity of the dripline of trees, where the most feeder roots are located. Aerating near the trunk should be avoided, due to the risk of puncturing large anchor roots. The recommended piece of equipment for aerating is a tine power aerator, which penetrates six to eight inches and pulls out narrow plugs of soil as it creates tubular holes. The soil plugs are left on the surface, where they breakdown over time. Power aerators can be rented or purchased as walk-behind or rider models. These types are preferred over tractor attachments, which are less maneuverable around the low-hanging perimeter of tree canopies. Each tree should receive an aerated band of orchard floor around the perimeter of the dripline, ranging from five to ten feet in width, depending on the size of the canopy. The aerated band is centered on the edge of the dripline, therefore aerating both inside and outside the canopy. Aerating immediately improves air and water penetration, and stimulates microbial activity and root development. For optimal benefits, aerating is followed by nutritional mulching.

Mulching

Mulching is recommended as a stabilization activity as it relieves health stressors, provides multiple benefits, and facilitates mowing, an important stabilization activity. Mulching involves the application of a layer of organic material beneath the canopy of each fruit tree, beginning just outside the dripline and reaching close to the trunk. The layer of mulch slowly breaks down over time, returning organic matter and nutrients to the soil, and stimulating microbial activity, leading to more tree root development and disease resilience. Mulch also suppresses water evaporation and the growth of ground cover vegetation, alleviating the need to irrigate and mow in the mulched area. The recommended type of mulch is finely shredded bark mulch with added nutrients, such as mushroom or chicken compost. The breakdown of mulch is facilitated by soil temperatures above 50 degrees Fahrenheit, and organic matter is more readily conducted to the root zone when soil aeration holes are present. A two to three inch-deep layer of mulch is typical, though care should be taken to hold the mulch 2 inches away from the trunk of the tree, creating a gap. Allowing the mulch to accumulate against the tree trunk can invite pest and disease problems.

The recommended equipment for mulching is a mulching fork (which looks rather like a widely-tined pitch fork), with mulch delivered to local stockpiles in a dump truck. After spreading with the fork, the mulch is leveled with a rake to establish an even depth.

Figure 50: A stable historic orchard in Olympic National Park has received deadwooding, aerating, and then mulch within the driplines of trees. In this case, the stabilization work was performed by an arborist contractor (NPS 2009).

Once mulch is delivered to the site, mulching is a suitable activity for trained volunteers, as no power tools are required. Where old fruit trees have not yet been deadwooded or "cleaned", head protection is recommended for the mulching crew. Ideally, however, deadwood removal is performed before mulching. Finely shredded bark mulch with added compost will gradually breakdown over a period of two to three years. Mulching should be repeated every two years during the stabilization period and will result in improvements to the condition class of trees. Mulching can be performed at any time of year, though moist soil conditions are most favorable.

Sucker, Watersprout and Crossing Branch Removal

A limited range of pruning is included within the scope of stabilization. The full repertoire of pruning is used in preservation maintenance, which is beyond the scope of stabilization. Preservation maintenance retains good conditions over time, either as a preservation treatment or after the implementation of a restoration or rehabilitation treatment. The preservation maintenance pruning of old fruit trees includes winter pruning to stimulate vigor by heading back or thinning the canopy, and summer pruning to reduce vigor by removing water sprouts.

Stabilization pruning activities are performed upon trees that are in fair or poor condition, with the objective of preventing further deterioration and improving condition. The range of pruning activities that are within the scope of stabilization includes just the removal of deadwood, root suckers, water sprouts and crossing branches. The removal of deadwood is addressed earlier in the document. This work can be performed at any time of year, and is usually one of the highest needs in stabilizing old fruit trees and orchards. Deadwooding removes a significant health stressor from the trees, but also creates safer working conditions around the trees for subsequent actions.

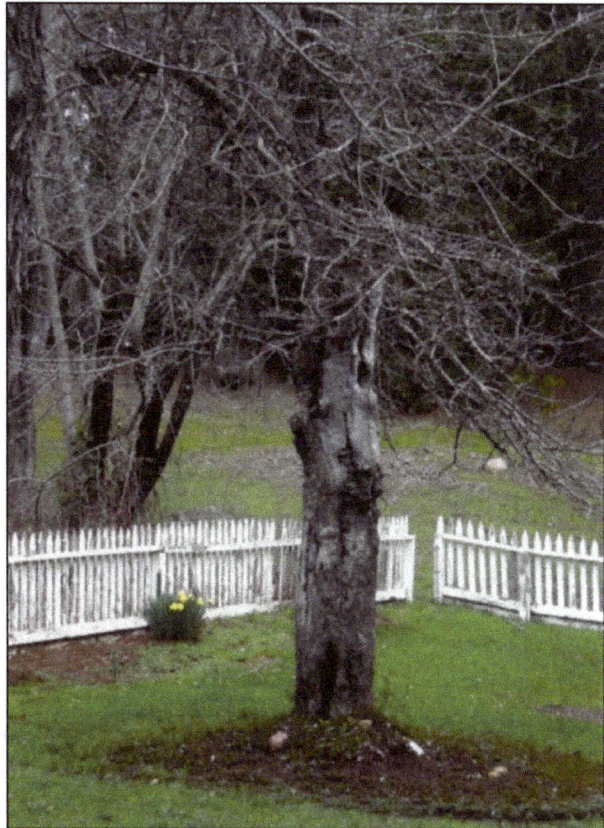

Figure 51: The canopy of this historic apple tree at Malakoff Diggings State Historic Park is a thicket of crossing branches (CDPR, 2010).

Root sucker removal is a suitable activity for volunteers, as it can be done without power tools from ground level. Root suckers emanate from the roots or rootstock of the tree, and very often emerge from the graft union at the base of the trunk. The presence of root suckers indicates vigor in the rootstock part of the tree, and some rootstocks are more vigorous and will produce more suckers, than others. If left unchecked, root suckers will grow up and develop into a new canopy engulfing the original canopy, posing competition for light, water and nutrients. Root suckers should be removed as soon as they appear at any time of year, but are more easily removed in the dormant season, when the leaves are off the tree and the root suckers, allowing greater visual access to the trunk. Occasionally, root suckers are temporarily allowed to remain on old fruit trees that are highly subject to browse predation by elk or deer. In this case, the thicket of root suckers is allowed to remain as a protective barrier around the trunk, making access for browsing on the canopy more difficult. This is a temporary measure until an acceptable fencing alternative can be implemented. More information on fencing is provided in the "Protection from Wildlife" section.

Root suckers are removed with a flush cut against the base of the trunk or the orchard floor, using a variety of hand tools. Finger-sized diameter root suckers can be removed with a pruning knife or a pair of pruners, while thick suckers should be removed with a pair of loppers or a handsaw.

Figure 52: A historic pear tree with severe root sucker development at Manzanar National Historic Site. Suckers are removable with hand tools (NPS, 2005).

All hand tools should be sterilized with isopropyl alcohol before the first cut, and then in-between successive trees. All debris should be removed from the site to promote sanitary conditions.

Watersprouts are vertical, unbranched vegetative shoots in the canopy that arise from the scion (or variety) part of the tree, unlike root suckers. Watersprouts are so-called as they "steal" water from the tree, while providing no fruit and contributing no structure to the scaffold as they are vertical and weak in bearing weight. The presence of water sprouts is indicative of vegetative vigor in the scion. Their removal is a stabilization action that checks and redirects this vigor within the scion to the shoots and spurs that are a desirable part of the permanent scaffold of the tree.

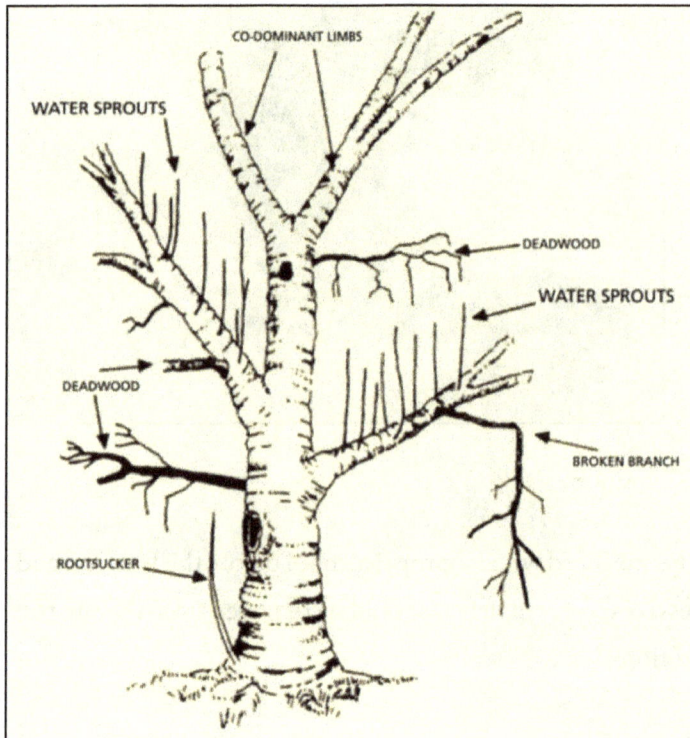

Figure 53: Diagram showing water sprouts and other tree problems that should be addressed during stabilization pruning (NPS, 2011).

The removal of watersprouts checks the vigor of the tree, and is therefore performed as part of summer pruning during the height of the growing season. Watersprouts are removed with a flush cut against the intersecting limb, using a variety of hand tools. Finger-diameter watersprouts can be removed with a pruning knife or a pair of pruners, while thicker watersprouts (older than one year) should be removed with a pair of loppers or a handsaw. All hand tools should be sterilized with isopropyl alcohol before the first cut, and in-between successive trees. All debris should be removed from the site to promote sanitary conditions.

Crossing branch removal rounds out the range of stabilization pruning activities. Crossing branches are two or more branches that converge rather than diverge. Pruning is necessary for stabilization when crossing branches are touching and abrading against each another, causing loss of bark and the potential for pest or disease entry. Stabilization's scope is limited to just removing the shortest length of limb possible to prevent further abrasion. In preservation maintenance, trees are maintained in good condition by winter pruning to ensure all branches grow directionally to the outside of the tree, and occupy their own physical space, rather like the spokes of a wheel. Neglected fruit trees often develop a thicket of inward-trending branches that will eventually cause abrasion problems. The scope of stabilization is limited to just stopping abrasion at the source rather than removing all inward-trending branches to restructure the canopy. The latter type of pruning will stimulate the vigor of the tree, and must be sustained by subsequent winter pruning efforts to maintain the structure.

Crossing branch removal is a skilled activity that should be performed by trained staff. Removal of limbs smaller than three inches in diameter is done using handsaws, loppers or pruners, and may be appropriate for trained volunteers. Larger limbs should be removed by a qualified chainsaw operator familiar with the following arborist's standards: ANSI A300 (Part 1) – 2001, "Tree, Shrub and Other Woody Plant Maintenance – Standard Practices (Pruning)". One of the crossing and abrading limbs is shortened to a node or axillary bud using a reduction cut, to prevent further abrasion. Reduction cuts are illustrated in the following figures. The limb to be shortened is selected based on various factors, including the health, form and location. Pruning equipment should be sterilized using isopropyl alcohol before the first cut, and in-between successive trees. Pruned debris should be removed from the site to maintain sanitary conditions.

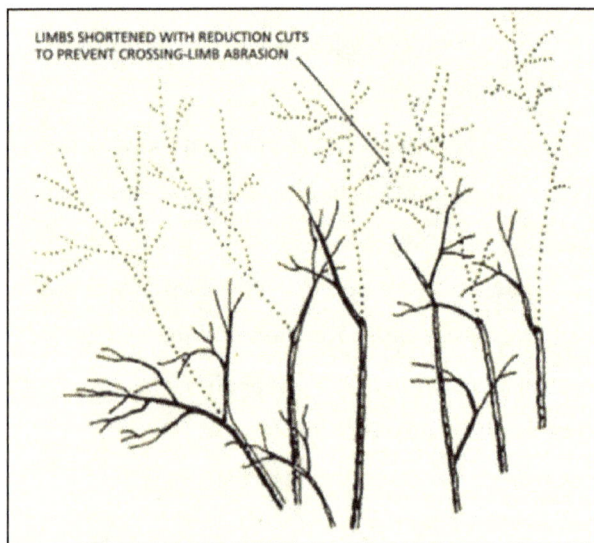

Figure 54: Limb shortening with reduction cuts to remove crossing branches and prevent abrasion. Cuts are made at an axillary bud (NPS, 2011).

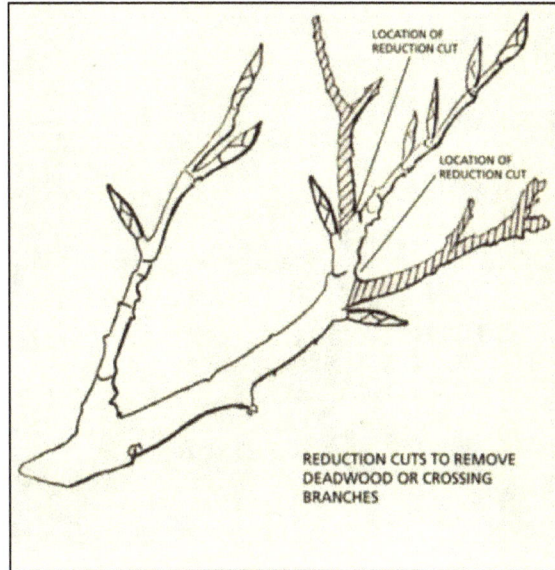

Figure 55: Magnified view of fine reduction cuts to remove deadwood or crossing limbs (NPS, 2011).

Fire Protection

Lightning and human-ignited wildfires are common in parts of California with potentially historic fruit trees and orchards. Whether slow or fast-moving, relatively hot or cool, wildfires pose a threat to old fruit trees as these species are not adapted to tolerate the effects of fire. Fire protection for fruit trees and orchards should be provided as part of stabilization efforts through fuels reduction on an ongoing basis, and the irrigation of trees when wildfire risk is high.

Figure 56: Severe wildfire damage to historic fruit trees in the North Coast Redwoods District (CDPR, 2010).

Fuels reduction alters the speed and temperature of fires; irrigation reduces the probability of ignition or spread. Fire protection by fuels reduction can be implemented by following the earlier guidance on "Encroaching Competitive Vegetation" and "Deadwooding". After removing encroaching competitive vegetation, the orchard floor should consist of herbaceous, non-woody plants that are maintained at a height of six inches or less. This condition promotes the vitality of old fruit trees but also drastically reduces fuels available to feed wildfires. A low-growing, herbaceous orchard floor stimulates the wildfire to move more quickly through a site with less heat, reducing the damage to fruit trees.

Figure 57: A relatively fast moving, cooler temperature wildfire in this olive orchard at Folsom Lake State Recreation Area has not killed the trees (CDPR, 2010).

Deadwooding and debris removal also promote a fast moving, cool and less damaging fire when wildfires attack. The presence of debris and deadwood provides ladder fuels, giving fire a path to spread from the orchard floor into fruit tree canopies. Also, wildfires can more easily spread from an adjacent site when the fruit trees are surrounded by woody vegetation containing deadwood and dry debris. Vegetation management for fuels reduction should extend into any woody vegetation or brush at the perimeter of the orchard, by removing deadwood and debris, and establishing a clear zone between perimeter vegetation and the fruit trees. Old fruit trees or orchards with clean canopies, an herbaceous orchard floor and a clear zone between the fruit trees and perimeter vegetation, are more protected from devastating wildfires.

When an orchard of high cultural resource value is threatened by a devastating wildfire, additional protection can be provided by wildfire retardants. Generally, fire retardants are dropped from an aircraft or applied by ground crews around a wildfire's edges in an effort to contain its spread. Class A foams may be sprayed on brush or grass around the perimeter of an orchard site to create a fire break ahead of a wildfire. Fire retardant slurries may be dropped from an aircraft to prevent ignition. Fire retardants should not be directly applied to fruit trees, as they can eventually become toxic to the trees. Forest fire retardants are generally considered to be non-toxic, but even low-toxicity compounds carry some risk when fruit trees and other organisms are exposed to large amounts.

Figure 58: Accumulated debris provides fuel for wildfires near a historic orchard at Manzanar National Historic Site. All debris should be removed off site to reduce fire danger (NPS, 2005).

Fire retardants used in firefighting can be toxic to fish, wildlife and firefighters and their use within 300 feet of a water body is generally prohibited. The United States Forest Service is the federal agency that governs research and monitoring of the effect of fire retardants on biotic systems. In California, all fire retardants must be approved for use through registration with the California State Fire Marshal. In lieu of fire retardants, deadwood removal, orchard floor management and clear zone maintenance are the best management practices for wildfire protection.

Irrigation

Irrigation is included within the scope of stabilization efforts when severe drought is a deficiency that threatens the mortality of fruit trees. Severe drought is seen in desiccated foliage throughout the tree canopy and premature leaf drop. While most old fruit trees and orchards in California state parks are un-irrigated and experience a period of summer drought, not all of these trees are threatened. This is because old fruit trees have extensive root systems with taproots that seek and find deep reserves of ground water. Old fruit trees adapt to the abundance or scarcity of resources in their environment by limiting their live canopy size and their amount of annual growth.

Old fruit trees have typically developed a balance with their environment over time. This balance can easily be disturbed by creating large, abrupt changes to the amount of available resources, such as caused by heavy irrigation or fertilization. As soon as more inputs are provided, trees will try to respond with more outputs. The likely outputs are more vegetative growth and more fruit. While these may seem desirable, they can actually threaten the vitality of the tree if not supported by adequate pruning, fruit thinning, harvesting or tree bracing, to prevent excessive growth, weight bearing and leaning. Even more harmful is a sudden increase in the amount of inputs, but the increase is not sustained. A rapid change after a new equilibrium is reached can impose a fatal shock to an old fruit tree. Managers must carefully consider the balance the fruit tree has created with its environment, and avoid changes that will result in increased outputs, unless the new inputs are sustained. In stabilization, irrigation should be supplied only to prevent further loss from desiccation. Fertilization should be supplied only in a slow-release form such as nutritional mulch, unless a severe macro or micro nutrient deficiency is detected by a soil test. Gradual transitions facilitate tree adaptation and a new equilibrium, rather than abrupt changes.

If drought is threatening fruit trees, irrigation should be supplied in a slow and sustained manner. A drip irrigation system is the recommended method for delivering water to threatened fruit trees. A drip system is composed of rubber or poly tubing that is laid out on the surface of the orchard floor around the dripline of each fruit tree. Micro-emitters, either in-line or separately attached by feeder lines to the tubing, deliver a constant flow of water to the root zone in small droplets. Drip systems are typically fed by a pump at a well or a water tank.

A solar-powered pump is optimal for remote locations. A water tank should be located at the highest elevation of the site, to allow for gravity flow to the emitters.

Figure 59: A drip irrigation system is laid out around the driplines of historic pear trees at Manzanar National Historic Site (NPS, 2010).

The water tank is refilled by a water truck or with a hose from a water main. The size of the pump, pipes, tubing and emitters must be custom-designed for the site, to respond to the volume and pressure of water, along with length of run, gradient and therefore drop, in elevation. Many irrigation parts supply companies are able to provide design services and guidance with installation. The drip irrigation system should be controlled by a timer that is programmed to turn on at night, to reduce evaporation and loss of water. Trees severely in need of water should receive at least 12 hours of irrigation per week through the drip system, to allow for deep penetration. Once started, irrigation should be sustained until precipitation resumes. Use of a drip irrigation requires frequent inspections for wildlife puncture damage and the potential need for frequent repairs. Spare tubing and emitters should be kept on hand to expedite repairs and prevent water waste or tree desiccation.

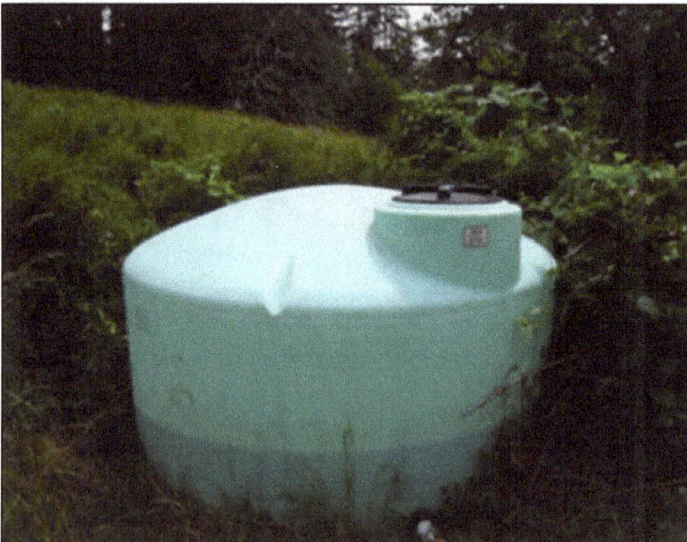

Figure 60: A water tank that supplies a drip irrigation system. The tank is located higher in elevation than the drip system (NPS, 2009).

Other methods for supplying irrigation water to stabilize old fruit trees and orchards include hand-watering and tree bladders. Hand watering usually requires a small tanker trunk or a pick-up truck with a tank or water bladder in the rear to be driven up and down the rows of trees. To promote deep penetration, a basin should be formed around the base of each tree by creating a rim of soil or mulch at the dripline to contain the water. Mulch is recommended to retain water in the root zone and retard evaporation. Mulch should not be allowed to touch the bark of the tree trunk, to discourage pests and diseases. In hand watering, the equivalent volume of one inch depth of water over the entire circle area within the dripline of the tree should be applied each week as slowly as possible, until precipitation resumes.

Tree bladders are semi-permeable bags that are placed on the orchard floor around the dripline of the tree, allowing for slow release of water into the ground. Tree bladders must be refilled on a weekly basis, involving the use of a small tanker truck or pickup truck with a water tank. There should be a sufficient number of bladders to supply the whole perimeter of the root zone with water. Filling the individual bladders is more time consuming than hand watering trees, however, irrigation is released more slowly and therefore effectively from the bladders. Once watering with tree bladders is initiated, irrigation should be sustained until precipitation resumes.

Figure 61: A basin within the dripline of a historic pear tree to facilitate hand watering. The efficiency could be improved with mulch (NPS, 2010).

Protection from Wildlife

Protection from browsing wildlife is part of the scope of stabilization if damage threatens the vitality of fruit trees. Potentially life-threatening damage to mature fruit trees from wildlife includes the loss of bark from the trunk and limbs, loss of limbs, and damage to the major roots. Damage to the fruits alone does not threaten the vitality of a fruit tree, and damage to foliage is not a life-threatening condition unless the tree has very little live canopy remaining. If serious damage is occurring, wildlife protection should be addressed within the scope of stabilization.

Figure 62: Potentially life-threatening loss of bark on a historic apple tree caused by winter browsing of elk (NPS, 2008).

Historically, browse protection was provided by the physical barrier of the orchard enclosure, by animal control, or by the deterrence posed by human presence. In the past, California state parks orchards were not the quiet, contemplative places they are today. Instead, fruit trees received frequent attention by their owners or workers tending to them or the surrounding land all year round. Wildlife would have been aware of the human presence, and would give the fruit trees a wide berth. In the present day, without regular tending for fruit production, fruit trees and orchards have become colonized by a diversity of wildlife that make use of the physical structure, shelter, and food resources of the orchard. Elk, deer, and bear reach into tree canopies, rubbing or tearing off the bark, breaking limbs, and toppling unsound trees. Ground squirrels, voles and other rodents strip bark and feed off root tissue. Many bird species build nests in

Figure 63: A pack rat nest in the canopy of a historic apple tree at Jack London State Historic Park. Pack rats feed on the shoots and fruits of the canopy and foster decay in limbs and trunks. Personal protective equipment must be worn when removing nests, due to the risk of Hanta Virus (NPS, 2008).

canopies or in the hollow cavities of trunks. Pack rats build nests in the crotch of scaffold limbs or in trunk cavities, introducing nutrients, persistent moisture and decay.

Browse protection strategies include physical barriers, chemical repellants and live trapping. Each strategy poses some challenges, whether due to visual intrusion, cost, labor-intensity, wildlife interactions or cultural or natural resources protection. The strategy or combination of strategies must be selected on the basis of multiple management objectives, including the type of damage to be addressed, resources protection, visitor experience and safety, funding and staffing capacities. Once implemented, the protection strategy must be sustained in order to be effective.

Physical barriers include fences, cages, and tree guards. Adequately designed and well-maintained fencing is the only failsafe tree protection from deer and elk browsing. However, fencing may be a large capital investment and an alteration of the site with the potential to have a visual impact on landscape character and the visitor experience. In historic sites or districts, fencing should meet the rehabilitation requirements of the *Secretary of the Interior's Standards for the Treatment of Historic Properties,* as well as meet deer or elk-proofing specifications. Fences meeting the Standards are compatible yet distinct from the historic features of the historic site or district.

Figure 64: A NPS Orchardist celebrates the completion of an elk and bear fence to protect the historic Buckner Orchard in North Cascades National Park Service Complex from further browse damage (NPS, 2009).

Depending on the species of deer or elk present and the maximum snow depth in winter, fencing may range from five feet in height (for smaller deer and elk, such as Tule), to eight feet on level ground. Deer will not typically jump a six-foot fence, but if chased or threatened, a large deer can clear eight feet on level ground. On sloping ground, a 10 to 11-foot fence may be needed to guard against deer or elk jumping down slope. Regardless of the material used to infill the fence, the design should have no spaces larger than six inches by six inches, and the fence should be fitted tightly to the ground. Deer and elk are more likely to crawl under a fence than attempt to jump over it. If wire is a compatible design material in the landscape, nine or ten gauge should be sufficiently strong. Fence posts must be installed at intervals sufficient to maintain tension in the lower and upper fence. A post interval of no greater than eight feet is recommended and posts should be well-anchored to prevent lateral movement.

Gates must also be considered in the design of a fence, equaling the height of the fence and the infill must be tight to the ground. Where a park orchard has a historic circulation system of roads or trails, gates should be located to preserve the use of the historic circulation system, rather than create a new one. Gates should permit equipment entry for maintaining the orchard floor, and should incorporate universal design features for visitors with disabilities.

Cages or individual tree fences are an alternative to fencing whole orchards, in order to provide protection from deer or elk. Fencing individual trees may be more practical or economical than fencing an entire area and this eliminates the need for large equipment gates. Individual cages also allow browsing animals to access the site, enabling grazing of the orchard floor with benefits for vegetation management. Like fences, where cages are an addition to a historic site or district, they have the potential to alter the historic character and the visitor experience. They may in fact have a greater visual impact than a singular perimeter fence. They also pose more challenges for maintenance access. Cages are generally temporary installations, however, and are easily removed when scaffold limbs and canopy are stabilized well above the browse height. Their design should be compatible with the historic character of the landscape, meeting the rehabilitation requirements of the *Secretary of the Interior's Standards*. They should meet the same animal-proofing specifications as fences. In designing cages, accessible openings with locking mechanisms must be included for maintenance access, and cages should allow space for working on each tree. A mulched floor inside the cage alleviates the need for mowing on the interior.

Figure 65: An example of temporary tree cages in a historic orchard at San Juan Island National Historical Park. In this case, the welded wire cages are being used to protect new trees that were planted as part of a restoration effort (NPS, 2008).

Tree guards are made of hard plastic or PVC, and are wrapped in spirals around the vulnerable trunk, beginning one inch below the soil, if possible. Guards have pores to allow for the exchange of heat and moisture, and pose a physical barrier to small mammals such as squirrels, rabbits, mice and voles. Tree guards should not be confused with tree wraps that protect immature trees from sun scald, which causes young, thin bark to split. Wraps are composed of paper or plastic tape, and are not needed in old fruit trees with mature bark development. Tree guards are generally not appropriate for use in extremely hot climates, where they can cause excessive heat accumulation and damage to the tree trunk.

Figure 66: Apple trees in the historic Buckner Orchard bear rigid plastic tree guards for protection against rodent damage (NPS, 2009).

Chemical repellants can be used to treat individual fruit trees that for various reasons, including resource protection, visitor experience or operational constraints, cannot be fenced or caged. Here the use of chemical repellant such as "Liquid Fence" is recommended. Chemical repellants have limited viability however, and need to be reapplied on a weekly basis in wet weather or as new foliage develops in order to be effective. The trunk and the browse-threatened shoots and foliage of the tree are coated with repellant spray. Wildlife is repelled by the odor or taste of the spray, and move on in search of other forage. Most repellants are not permitted on food crops, and so make sure the repellant is registered for this use, if fruit production is a management objective. When using repellants, the product label directions should be followed carefully.

Live trapping is generally the environmentally preferred method of rodent control for ground squirrels, pack rats and voles. Fumigants or poisons can leave residues in the environment or be conveyed through the food chain. Multiple live traps are needed per orchard area. Traps baited with nuts, grains or fruits are placed near burrows or nests, and are checked on a daily basis. Carcasses should be removed with a gloved hand, as rodents can carry diseases harmful to humans. Live rodents caught in traps can be relocated or dispatched by fumigation.

Figure 67: A live trap at the mouth of a burrow at the base of a historic pear tree at Manzanar National Historic Site. The trap is baited with nuts or grain and trapped animals are collected and dispatched (NPS, 2010).

Repairs Using Bridge Grafting

Bridge grafting is included within the scope of stabilization when fruit trees are threatened by the severe loss of bark. Bridge grafting is used to repair injuries to trunks or limbs in which large areas of bark have been destroyed. The damage may have been due to girdling by wildlife, inadvertent mechanical damage, or the presence of a cavity. A bridge of scion wood is grafted above and below the damaged area, allowing the conductive tissue to be reconnected. Rather like bypass surgery, bridge grafting allows a new pathway for the circulation of water and nutrients, effectively healing the wound and allowing for a return to stable conditions.

Bridge grafting is an old horticultural technique that was commonly practiced by orchardists in California and throughout the country in the 19th and early 20th-centuries. As with all types of grafting, bridge grafting is a highly skilled technique, and should only be performed by an experienced orchardist or a trained arborist. The bridge material is obtained from one-year old dormant scion wood of the damaged tree, or another suitable tree of the same species and variety, if possible. The scion wood is kept dormant by refrigeration until needed for use. In early spring, just as dormancy is beginning to break and the "bark slips against the cambium", the damaged area is prepared for grafting by cleaning away dead debris and jagged tissue. The scion wood is cut to length to bridge the wound, allowing some length for flexibility, and the ends are precisely cut in a wedge profile, using a grafting knife. Each side of the damaged area is prepared with small slits in the bark to create a flap to expose the cambium. With the scion wood positioned vertically (i.e., the axillary buds are oriented upwards), the upper and lower ends are slipped into the flaps, allowing the wedge ends of the scion wood to touch the exposed cambium. Each end of the bridge is secured using stainless steel brads and grafting wax.

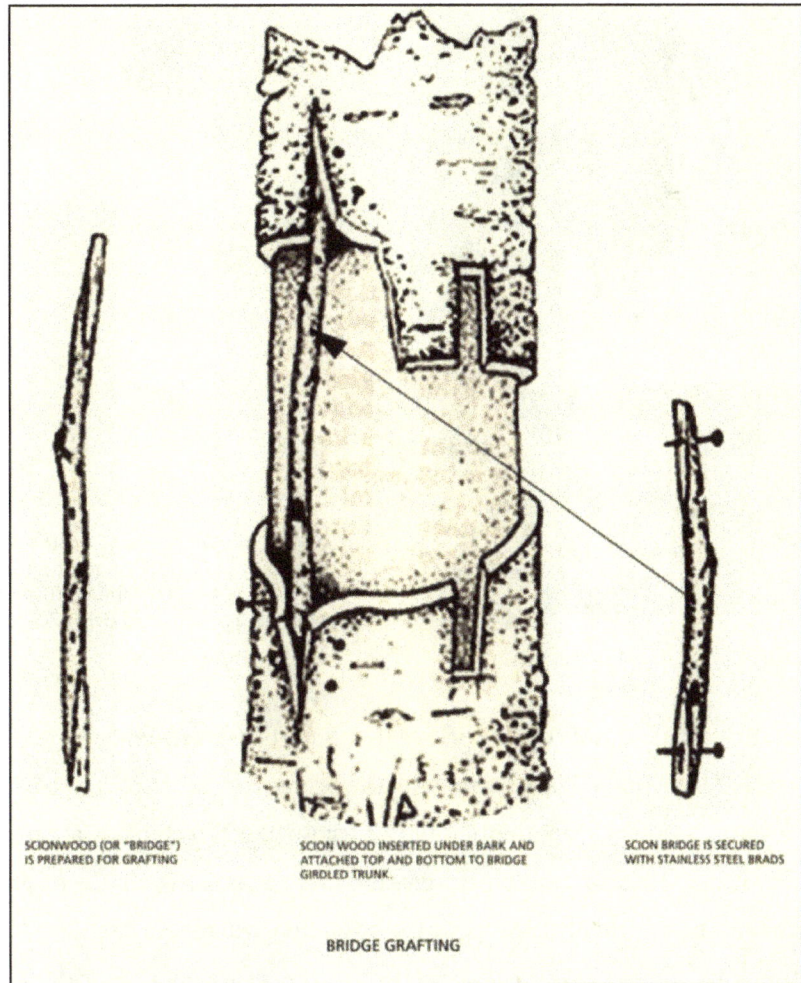

SCIONWOOD (OR "BRIDGE")
IS PREPARED FOR GRAFTING

SCION WOOD INSERTED UNDER BARK AND
ATTACHED TOP AND BOTTOM TO BRIDGE
GIRDLED TRUNK.

SCION BRIDGE IS SECURED
WITH STAINLESS STEEL BRADS

BRIDGE GRAFTING

Figure 68: Diagram of a bridge graft using scion wood from the canopy of the damaged tree to repair a girdled area. The scion wood bridge is slipped under the bark flap and is secured with stainless steel brads (NPS, 2011).

Figure 69: A year after bridge grafting, the multiple bridges in this repair have begun to grow and will eventually expand around the wound, effectively sealing over the heartwood (NPS, 2011).

In the year after grafting, the bridge tissue starts to swell, and eventually will expand in girth. Axillary buds along the bridge may attempt to break bud and develop into stems, but they should be trimmed off to concentrate the energy in the bridge. Usually multiple bridges are grafted, as shown in the photograph above, and eventually, as the bridges swell they will merge laterally, sealing over the heartwood or cavity on the interior. Note, however, that bridge grafts are not usually 100% successful, and may need to be repeated with more dormant scion wood in the next spring.

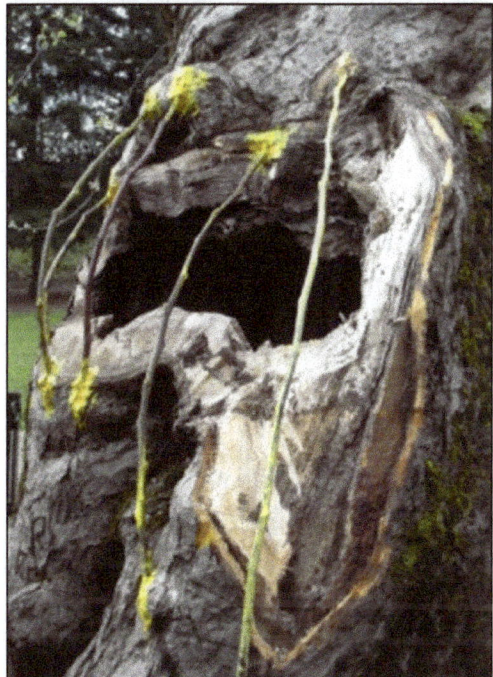

Figure 70: Photos illustrating the bridge grafting of a 185-year old apple tree at Vancouver Reserve in Washington, showing a close-up view of the bark flap and inserted bridge (upper photo), and completed repairs secured with yellow grafting wax (lower photo) (NPS, 2009).

APPENDICES

California State Parks With Fruit Trees More Than Fifty Years Old (Potentially Historic)

In 2008, a system-wide survey of California state parks identified the following parks a having fruit trees that are potentially more than fifty years old. The survey was conducted via the internet using an online survey tool. The voluntary responses by district staff yielded results indicating that a minimum of 44 parks have potentially historic orchards or fruit trees.

Anderson Marsh SHP

Andrew Molera SP

Auburn SRA

Benicia Capitol SHP

Bidwell-Sacramento River SP

Bothe-Napa Valley SP

Burleigh H. Murray Ranch Property

California Citrus SHP

Castle Crags SP

Castro Adobe Property

Clear Lake SP

Del Norte Coast Redwoods SP

Delta Meadows RP (Locke Boardinghouse)

Folsom Lake SRA

Fort Ross SHP

Henry W. Coe SP

Hollister Hills SVRA

Humboldt Lagoons SP

Humboldt Redwoods SP

Hungry Valley SVRA

Jedediah Smith Redwoods SP

Lake Oroville SRA

Lake Perris SRA

Los Encinos SHP

Malibu Creek SP

Marshall Gold Discovery SHP

McArthur-Burney Falls Memorial SP

Monterey SHP

Old Town San Diego SHP

Olompali SHP

Palomar Mountain SP

Pio Pico State Historic Park

Prairie Creek Redwoods SP

Robert Louis Stevenson SP

San Juan Bautista SHP

Silverwood Lake SRA

Sinkyone Wilderness SP

Shasta SHP

Sonoma SHP

Sutter Buttes SP

Tolowa Dunes State Park

Topanga SP

Will Rogers SHP

CDPR Orchards and Fruit Trees 2008 Survey Responses

No.	Park	District			Potentially Non-...less	Potentially Non-...
1	Admiral William Stanley SRA	North Coast Redwoods	N	N/A	N	N/A
2	Ahjumawi Lava Springs SP	Northern Buttes	Y	Apple	N	N/A
3	Anderson Marsh SHP	Northern Buttes	Y	Walnut	Y	Walnut
4	Andrew Molera SP	Monterey	Y	Apple, Pear	N	N/A
5	Auburn SRA	Gold Fields	Y	Apple, Peach, Plum, Pear, Fig, Olive, Persimmon, Walnut	Y	Plum, Fig, Walnut
6	Benicia Capitol SHP	Diablo Vista	Y	Orange, Fig, Olive, Walnut	N	N/A
7	Bidwell Mansion SHP	Northern Buttes	N	N/A	Y	Persimmon, Pecan
8	Bidwell-Sacramento River SP	Northern Buttes	Y	Walnut	Y	Walnut
9	Bothe-Napa Valley SP	Diablo Vista	Y	Apple	Y	Apple, Plum

No.	Park	District			Potentially Non-	ion-
10	Burleigh H. Murray Ranch Property	Santa Cruz	Y	Apple	N	N/A
11	California Citrus SHP	Inland Empire	Y	Orange, Lemmon, Olive, Walnut	Y	Orange, Lemon
12	Castle Craggs SP	Northern Buttes	Y	Apple	Y	Apple
13	Castro Adobe Property	Santa Cruz	Y	Apple, Pear	N	N/A
14	Clear Lake SP	Northern Buttes	Y	Apple	Y	Apple
15	Colonel Allensworth SHP	Tehachapi	N	N/A	Y	Pistachio Apple, Pomegranate
16	Colusa-Sacramento River SRA	Northern Buttes	N	N/A	Y	Pomegranate
17	Cuyamaca Rancho	Colorado Desert	N	N/A	Y	Apple
18	Del Norte Redwoods SP	North Coast Redwoods	N	N/A	N	N/A
19	Delta Meadows SP	Gold Fields	Y	Pear, Fig, Walnut	Y	Plum, Fig, Walnut
20	Folsom Lake SRA	Gold Fields	Y	Orange, Apple, Lemon, Plum, Pear, Nectarine, Fig, Olive, Persimmon, Walnut	Y	Plum, Fig

No.	Park	District			Potentially Non-	Potentially Non-Fruit Trees
2 1	Fort Ross SHP	Russian River	Y	Apple (Russian Bellflower, Gravenstein), Plum, Pear (Vicar of Wakefield, Bartlett), Bitter Cherry, Fig, Olive	Y	Apple
2 2	Fort Tejon SHP	Tehachapi	N	N/A	Y	Apple, Peach, Plum, Pistachio Pomegranate
2 3	Henry W. Coe SP	Monterey	N	N/A	N	N/A
2 4	Hollister Hills SVRA	OHMR	Y	Olive, Walnut	Y	Walnut
2 5	Humboldt Lagoons SP	North Coast Redwoods	Y	Apple	N	N/A
2 6	Humboldt Redwoods SP	North Coast Redwoods	Y	Apple, Plum, Pear, Cherry, Quince, Fig, Walnut	Y	Unidentified
2 7	Hungry Valley SVRA	OHMR	Y	Almond, Black Walnut	Y	Almond, Peach, Plum, Walnut

No.	Park	District			Potentially Non-Historic	Potentially Non-Historic Species
28	Jedediah Smith Redwoods SP	North Coast Redwoods	Y	Apple	N	N/A
29	John Marsh Home SHP	Diablo Vista	N	N/A	Y	Walnut
30	La Purisima Mission SHP	Channel Coast	N	N/A	N	N/A
31	Lake Oroville SRA	Northern Buttes	Y	Olive, Walnut, Orange	N	N/A
32	Lake Perris SRA	Inland Empire	Y	Pear, Pomegranate, Quince, Fig, Olive,	N	N/A
33	Los Encinos SHP	Angeles	Y	Orange, Olive	Y	Unidentified
34	Malakoff Diggings SHP	Sierra	N	N/A	N	N/A
35	Malibu Creek SP	Angeles	Y	Pomegranate, Persimmon, Mandarin Orange, Olive, Lime, Pineapple Guava, Kumquat	N	N/A
36	Marshall Gold Discovery SHP	Gold Fields	Y	Apple, Peach, Plum, Mulberry, Pear, Fig, Walnut	Y	Orange, Apple, Peach, Plum, Fig, Persimmon, Walnut

No.	Park	District	...istoric		Potentially Non-	Potentially Non-
3 7	McArthur-Burney Falls Memorial SP	Northern Buttes	Y	Apple	N	N/A
3 8	Monterey SHP	Monterey	Y	Apple, Apricot, Quince, Olive	N	N/A
3 9	Old Town San Diego SHP	San Diego Coast	Y	Pomegranate, Quince, Fig, Olive	Y	Orange, Apricot, Pomegranate, Quince, Fig
4 0	Olompali SHP	Marin	Y	Pomegranate	Y	Walnut
4 1	Palomar Mountain SP	Colorado Desert	Y	Apple	N	N/A
4 2	Pio Pico SHP	Angeles	N	N/A	Y	Unidentified
4 3	Prairie Creek Redwoods SP	North Coast Redwoods	Y	Apple	N	N/A
4 4	Robert Louis Stevenson SP	Diablo Vista	Y	Apple, Peach, Plum	N	N/A
4 5	San Juan Bautista SHP	Monterey	N	N/A	N	N/A
4 6	Silverwood Lake SRA	Tehachapi	Y	Apple, Plum, Pear	N	N/A
4 7	Sinkyone Wilderness SP	North Coast Redwoods	Y	Apple	N	N/A

No.	Park	District			Non-	Potentially Non-
48	Shasta SHP	Northern Buttes	Y	Unidentified	Y	Apple, Peach, Apricot, Pear, Cherry, Fig, Walnut
49	Sonoma SHP	Diablo Vista	Y	Orange, Pomegranate, Olive	Y	Quince
50	Sutter Buttes SP	Northern Buttes	Y	Almond	N	N/A
51	Tolowa Dunes SP	North Coast Redwoods	Y	Apple	N	N/A
52	Topanga SP	Angeles	Y	Orange	N	N/A
53	Will Rogers SHP	Angeles	Y	Cherimoya	N	N/A
54	Woodson Bridge SRA	Northern Buttes	N	N/A	Y	Plum, Apricot

Common Fruit Varieties in California Pre-World War II

Almond

Jordanolo

Ne Plus Ultra

Nonpareil

Texas

Apple (varieties most popular in 1910)

Baldwin

Ben Davis

Beverly Hills

Esopus Spitzenberg

Gravenstein

Grimes Golden

Jonathan

King of Tompkins County (Tompkins King)

Newtown Pippin (Yellow Newtown)

Northern Spy

Rhode Island Greening

Roxbury Russett

Starking

White Pearmain

Winesap

Winter Banana

York Imperial

Apple (varieties most popular in 1942)

Golden Delicious

Grimes Golden

Jonathan

McIntosh

Newtown Pippin

Northern Spy

Red Delicious

Rhode Island Greening

Rome Beauty

Stayman Winesap

York Imperial

Apricot

Blenheim

Royal

Tilton

Cherimoya

Booth

Deliciosa

Cherry

Bing

Black Tartarian

Black Republican

Burbank

Chapman

Lambert

Napoleon

Fig

Kadota

Mission

Turkey (Brown Turkey)

Lemon

Eureka

Lisbon

Nectarine

Gold Mine

John Rivers

Quetta

Victoria

Olive

Manzanillo

Mission

Sevillano or "Queen"

Orange

Navel

Robertson Navel

Valencia

Peach (Freestone varieties most popular in 1930)

Belle

Carman

Early Crawford

Elberta

J.H. Hale

Late Crawford

Levi

Lovell

Muir

Salwey

St. John

Triumph

Peach (Clingstone varieties **most popular in 1930**)

Babcock

Curlew

Early Wheeler

Golden Brush

Haus

Heath Cling

Meadow Gold

Orange

Palono

Peaks

Phillip

Robin

Sun Glow

Tuscan

Pear

Anjou

Bartlett

Bose

Comice

Hardy

Winter Nelis

Persimmon

Fuyu

Hachiya

Plum (European varieties (*Prunus domestica* and *P. institia*)
most popular in 1930)

Arch Duke

Bavay

Bradshaw

Damson

German Prune

Grand Duke

Green Gage (Reine Claude)

Imperial Epineuse

Italian Prune (Fellenberg)

Lombard

Monarch

Moore Arctic

Pacific

Pond

Quackenboss

Shropshire Damson

Tennant

Tragedy

Yellow Egg

Plum (Japanese varieties (*Prunus triloba*) most popular in 1930)

Abundance

Beauty

Burbank

Chabot

Climax

Duarte

Formosa

Gaviota

Kelsey

Inca

Mariposa

Red June

Santa Rosa

Satsuma

Wickson

Pomegranate

Wonderful

Prune

Burton

French

Sugar

Quince

Smyrna

English Walnut

Concord

Eureka

Franquette

Hartley

Placentia

Poe

Relevant Organizations

USDA, National Germplasm Resources Laboratory

10300 Baltimore Boulevard

Room 103, Building 003, BARC-West

Beltsville, Maryland 20705

USDA, Agricultural Research Service

National Plant Germplasm System

http://sun.ars-grin.gov/npgs

National Plant Germplasm Repository for Pear – Corvallis, Oregon

USDA, Agricultural Research Service

33447 SE Peoria Road

Corvallis, Oregon 97333-2521

National Plant Germplasm Repository for Citrus and Dates – Riverside, California

USDA, Agricultural Research Service

1060 Martin Luther King Boulevard

Riverside, California 92507-5437

National Plant Germplasm Repository for Tree Fruit/Nut Crops and Grapes – Davis, California

USDA, Agricultural Research Service

One Shields Avenue, UCD

Davis, California 95616-8607

National Plant Germplasm Repository for Subtropical Horticulture – Citrus

Subtropical Horticultural Research Station

USDA, Agricultural Research Service

13601 Old Cutler Road

Miami, Florida 33158

National Plant Germplasm Repository for Tropical Fruits and Nuts

USDA, Agricultural Research Service

P.O. Box 4487

928 Stainback Highway

Hilo, Hawaii 96720

National Collection of Genetic Resources for Pecan and Hickories

USDA, Agricultural Research Service

10200 FM 50

Somerville, TX 77879

National Plant Germplasm Repository for Apples

USDA, Agricultural Research Service

Plant Genetic Resources Unit

Cornell University Experiment Station

630 W. North Street

Geneva, New York 14456-0462

California Rare Fruit Growers, Inc.

The Fullerton Arboretum

California State University at Fullerton (CSUF)

P.O. Box 6850

Fullerton, CA 92834-6850

http://www.crfg.org

University of California Statewide IPM Program

http://www.ipm.ucdavis.edu

GLOSSARY

Aeration

The mechanical process of making holes in the orchard floor near the dripline of the fruit tree canopy, to increase air supply to the root zone.

Axillary bud

Small dormant lateral tissue which can "break forth" from a leaf axil and develop into a flower, twig or leaves.

Branch collar

An overlap of tissue of branch and branch, or branch and trunk, often appearing as a small bulge or collar, giving a ringed appearance.

Canopy

The branches and leaves altogether – the top of the tree (also known as the crown).

California Register of Historical Resources

A program designated by The State Historical Resources Commission for use by state and local agencies, private groups and citizens to evaluate, register and protect California's historical resources. The California Register is the authoritative guide to the state's significant historical and archaeological resources.

Certified Arborist

Tree service professional certified with the International Society of Arboriculture. Experience, tests, and continuing education are required for certification.

Clonal dwarfing rootstock

A commercial horticulture term for dwarfing rootstocks that are clones or genetically identical to each other. Dwarfing rootstocks are cloned to perpetuate desirable characteristics and to guarantee a rootstock's ability to confer these characteristics upon the scion, such as extent of dwarfness, disease resistance, and youthful bearing of fruit.

Clone

The scion portion of a tree propagated by grafting. A clone is genetically identical to the parent. Clones, as opposed to seedlings, do not have genetic variation (see also: "scion").

Crotch

Top of the union or merging of two branches, or branch and trunk, or two leaders.

Crown

Portion of the tree above ground comprised of all the branches and foliage.

Cultivar

The abbreviated term for a "cultivated variety." A man-made variation within a species. The name of the cultivar follows the genus and species and is denoted by single quotation marks. The initial letters of the cultivar name are capitalized, e.g., *Pyrus communis* 'Winter Bartlett,' and the cultivar name is not italicized (see also: "variety").

Cultural resource

Term for a building, site, district, object, or structure evaluated as historically significant. Sites or districts may also be composed of biotic cultural resources, such as historic plants, vegetation and animals.

Deadwood

Dead twigs, branches or scaffold limbs either attached or hanging within the canopy that should be removed along with diseased or damaged growth, as soon as it becomes evident.

Deciduous

Plants or trees that drop leaves, needles or foliage in winter.

Dieback

When ends of twigs or branches defoliate, decline and die back to remaining live tree parts. A totally dead tree has no dieback.

Dripline

The perimeter or boundary of the canopy at ground level – generally in a circular line. The majority of feeder roots are located at, just beyond, or just within, the dripline.

Dwarfing rootstock

A rootstock that limits the height of a grafted tree to be shorter than the standard height (see also: "dwarf tree").

Dwarf tree

A tree grown on a rootstock that limits its final height to be shorter than the standard height. Dwarf trees are generally classified as semi-standard, about two thirds of standard height; semi dwarf, about half of standard height; and dwarf, about one third of standard height.

Evaluation

National Register/ California Register term for the process by which the significance and integrity of a historic property are judged and eligibility for listing in the National Register of Historic Places and the California Register of Historical Resources is determined.

Germplasm

The genetic material, especially its specific molecular and chemical constitution that forms the physical basis of heredity and is transmitted from one generation to the next. When applied to plants, it is the term given to seed or any vegetative material from which plants can be propagated.

Grafting

A method of vegetative propagation in which two different plants are joined together in order to take advantage of the special characteristics of each (see also: "rootstock" and "scion").

Graft union

The joint between the two parts of the grafted tree which have grown together. When visible, the union appears as a line, scar, indent, or change in bark pattern on the tree trunk. The height of the graft union on the trunk has varied over time, and during the early 20th-century, at ground level (see also: "grafting").

Historic context

National Register/ California Register term for an organizing structure for interpreting history that groups information about historic properties which share a common theme, common geographical area, and a common time period. The development of historic contexts is a foundation for decisions about the planning, identification, evaluation, registration, and treatment of historic properties, based upon comparative historic significance.

Historic integrity

National Register/ California Register term for the unimpaired ability of a property to convey its historical significance. Integrity is a measure of the physical authenticity of a historic property or cultural resource.

Historic significance

National Register/ California Register term for the value or importance of a historic property within the patterns of American history, in relation to a historic context. Significance may be in association with important events or persons, or for importance in design or construction, or for information potential.

Historic vegetation

The term for the plants growing within a historic site or district that date from the period of significance, also known as biotic cultural resources. The plants may be native or non-native, introduced (i.e., planted) or naturally-occurring on the land. Historic vegetation is one characteristic of historic sites or districts that is preserved to retain the historic integrity of the property.

Low-headed tree

The term for a tree with scaffold limbs borne upon a short trunk. The head or point of attachment of the main branches to the trunk is set by pruning in the first or second year after planting. The practice of low heading or creating fruit trees with a low head on a short trunk, was used to control heights in the transition from standard to dwarf trees between 1881 and 1945.

National Register of Historic Places

The Nation's official list of cultural resources worthy of preservation. Authorized under the National Historic Preservation Act of 1966, the National Register is part of a national program to coordinate and support public and private efforts to identify, evaluate, and protect our historic and archeological resources. Properties listed in the Register include districts, sites, buildings, structures, and objects that are significant in American history, architecture, archeology, engineering, and culture. The National Register is administered by the National Park Service, which is part of the U.S. Department of the Interior.

Node

The point of attachment of leaves and axillary buds (the stem between the nodes is called an "internode").

Rootstock

The term used in grafting to refer to the root system (see also: "scion"). Grafted fruit trees are composed of two genetically unique individuals, rather like two trees in one. The rootstock is joined to the scion (composing the trunk and canopy of the trees). The rootstock confers growth characteristics but does not affect the type of variety. Variety is conferred by the scion.

Root sucker

A shoot originating from a root, or from a rootstock either at or below the level of the graft union with the scion. Root suckers should be removed, to avoid competition with the tree.

Scaffold

The framework of major branches or limbs growing from the trunk on a tree.

Scion

The term used in grafting to refer to the upper portion of the graft, typically the aerial portion of the grafted tree. Grafted trees are composed of two genetically unique individuals, rather like two trees in one. The scion is joined to the rootstock (the subterranean part of the tree). The scion has the genotype of the cultivated variety, influencing the type, color, flavor and qualities of fruit.

Seedling

A tree originating from a seed rather from vegetative propagation by grafting.

Soil Test

A chemical test of a soil sample that determines pH and nutrient content, including macro nutrients (Nitrogen, Phosphorus and Potassium) and micronutrients.

Stabilization

The set of interim actions applied to prevent the further deterioration of a potential or known cultural resource, and improve the condition from poor to fair or fair to good.

Standard tree

A tree grown on its own roots or grafted onto a seedling rootstock that allows the tree to reach its natural height (see also: "dwarf tree").

Variety

A naturally-occurring variation within a species. The variety name is a Latin name written after the genus and species. The variety name is italicized along with the genus and species, e.g., *Prunus cerasifera atropurpurea* (see also: "cultivar").

Vegetative propagation

The process of producing a new plant from a portion of another plant, such as a stem or a branch. Also known as asexual reproduction, the process does not involve the mixing of genes from different parents as in sexual reproduction. The new offspring is genetically identical or a clone of the parent. Grafting is a method of vegetative propagation.

Water sprout

A vertical shoot from a branch or upper trunk that is fast growing, vegetative and causes an excessively heavy or weak-wooded canopy, if allowed to remain. Generally, all water sprouts should be removed from the fruit tree during summer pruning.

www.ingramcontent.com/pod-product-compliance
Lightning Source LLC
Chambersburg PA
CBHW080553220326

41599CB00032B/6464